19.84

Universitext

Advisors
Paul R. Halmos   F.W. Gehring
C.C. Moore

D1452961

Paul Kelly
Gordon Matthews

The Non-Euclidean, Hyperbolic Plane

Its Structure and Consistency

With 201 Illustrations

Springer-Verlag
New York Heidelberg Berlin

Paul Kelly
Department of Mathematics
University of California
Santa Barbara, CA 93106
USA

Gordon Matthews
Department of Mathematics
California State University at
   Dominguez Hills
Carson, CA 90747
USA

AMS Subject Classifications (1980): 51-XX, 51-M10

Library of Congress Cataloging in Publication Data

Kelly, Paul Joseph.
   The non-Euclidean hyperbolic plane.

   (Universitext)
   Bibliography: p.
   Includes index.
   1. Geometry, Hyperbolic.   I.   Matthews, Gordon,
1915-   . II. Title.
QA685.K42   516.9   81-1389
                         AACR2

© 1981 by Springer-Verlag New York Inc.
Printed in the United States of America.

9 8 7 6 5 4 3 2 1

ISBN 0-387-90552-9 Springer-Verlag New York Heidelberg Berlin
ISBN 3-540-90552-9 Springer-Verlag Berlin Heidelberg New York

To
Pat and Betsey

Contents

Preface

The discovery of hyperbolic geometry, and the subsequent proof that this geometry is just as logical as Euclid's, had a profound influence on man's understanding of mathematics and the relation of mathematical geometry to the physical world. It is now possible, due in large part to axioms devised by George Birkhoff, to give an accurate, elementary development of hyperbolic plane geometry. Also, using the Poincaré model and inversive geometry, the equiconsistency of hyperbolic plane geometry and euclidean plane geometry can be proved without the use of any advanced mathematics. These two facts provided both the motivation and the two central themes of the present work.

Basic hyperbolic plane geometry, and the proof of its equal footing with euclidean plane geometry, is presented here in terms accessible to anyone with a good background in high school mathematics. The development, however, is especially directed to college students who may become secondary teachers. For that reason, the treatment is designed to emphasize those aspects of hyperbolic plane geometry which contribute to the skills, knowledge, and insights needed to teach euclidean geometry with some mastery.

The plan of the book can be sketched quite simply. Chapter I outlines the history of the "parallel-postulate" problem and concludes with an explanation of the distinction between absolute geometry and euclidean geometry. Chapter II is a review of the Birkhoff axioms and the principal theorems of euclidean plane geometry preceding the theory of parallel lines and similarity, i.e. absolute geometry. No proofs are given in the first three sections of this chapter, since the material is assumed to be familiar. In section 5, a basic circle property is derived to illustrate the use of a Dedekind cut. Section 6 introduces point and line reflections of the absolute plane and establishes a few of the basic invariants of such motions. While the material in this chapter can be reviewed rather quickly, familiarity with its content is essential since that content provides the foundation for hyperbolic plane geometry and is referred to frequently.

In Chapter III, a denial of the euclidean Playfair axiom is introduced as the characteristic hyperbolic axiom. With this axiom added to the system of absolute geometry, Chapter III presents a synthetic development of the central theorems in hyperbolic plane geometry. While the sequence of the theory is fairly traditional, more than usual care is taken with the accuracy of order relations, and the non-traditional use of mappings adds interest as well as simplicity to many of the proofs.

Chapter IV is devoted to the second theme of the book, the Poincaré model representation of hyperbolic geometry. A necessary background from the theory of circular inversions of the euclidean plane is developed independently. With this background, a systematic proof is given that the eight axioms of absolute geometry, as well as the hyperbolic axiom, are satisfied in the model geometry. Thus it is established that any inconsistency in hyperbolic plane geometry would have to imply a corresponding inconsistency in euclidean geometry.

Finally, a short introduction to distance geometries defined by a metric is given in an appendix. Its purpose is to provide a broader overview of non-euclidean geometries, and to give some indication of the nature of elliptic geometry, whose history is mentioned in the introductory chapter.

The first author has for several years taught a course along the lines of this book, and many non-mathematics majors have had success with the material. In his experience, the first three chapters can be adequately covered in a standard quarter course, and the entire four chapters in a semester course.

In conclusion, the authors wish to acknowledge their special indebtedness to Ernst Straus for his constructive criticism of the text and his many helpful suggestions.

Paul Kelly

Gordon Matthews

Principal Notations

1. The Absolute Plane:

A^2	the absolute plane
A,B,C	points
d(A,B)	distance between points A and B
R,S,T	sets of points
∅	the null set
r,s,t	lines; L(AB) line of A and B
S(AB), S[AB]	open and closed segments of A and B
S(AB], S[AB)	half open segments of A and B
R(AB), R[AB]	open and closed rays from A through B
⊁BAC	angle of rays R[AB)and R[AC).
⊁BAC°	degree measure of ⊁BAC
In(⊁BAC)	region interior to ⊁BAC
\<abc\>	the number b is between the numbers a and c
\<ABC\>	the point B is between the points A and C
⊥	perpendicularity symbol
∆ABC, In(∆ABC)	triangle and triangle interior
≅	congruence symbol
St(r,s)	open strip between non-intersecting lines r and s
C(A,r),	In[C A,r] ,Ex[C A,r] , circle with center A and radius r, interior and exterior to the circle.
Γ, Γ⁻¹	mapping and mapping inverse
$Γ_A$, $Γ_u$	reflection in point A, reflection in line u
P(A)	the pencil (family) of lines through point A.

2. The Hyperbolic Plane:

H^2	the hyperbolic plane
⊁(P,t)	the fan angle of point P and line t, P not on t
||	symbol of parallelism
)(	symbol of hyperparallelism
(B-AC-D)	biangle with ray sides R[AB), R[CD) and segment side S[AC]

In(B-AC-D)   interior to biangle (B-AC-D)

$P_1P_2\ldots.P_n$ proper, convex n-gon with successive vertices $P_1,P_2,\ldots P_n$

$\pi,\ \pi^{-1}$ the angle of parallelism function and its inverse

F[R(AB)]   the parallel family consisting of L(AB) and all lines parallel to R[AB)

F(b)   the hyperparallel family consisting of all lines perpendicular to base line b.

LC[R(AB);P]   the limit circle (horocycle) through P whose radial lines form the parallel family F[R(AB)].

EC(b;P)   the equidistant curve through P, not on b, whose radial lines form the hyperparallel family F(b)

In[LC[R(AB);P]], Ex[LC[R(AB);P]], regions interior and exterior to the limit circle

In[EC(b;P)], Ex[EC(b;P)], regions interior and exterior to the equidistant curve.

3.   The Euclidean Plane:

E^2 the euclidean plane

$\Phi_A(r)$ inversion in the euclidean circle C(A,r)

~   similarity symbol

Pw[P;C(B,b)] power of P with respect to the circle C(B,b)

F_p the family of lines and circle through point P

$F_p(s)$ the family of lines and circles through P which are tangent to the curve s in F_p

arc(PQ), arc[PQ]   open and closed arcs, of a circle with endpoints P and Q

CR(P,Q,R,S)   the cross ratio of the ordered quadruple (P,Q,R,S).

4.   The Poincaré Model:

H the space of points interior to the euclidean circle C(O,1)

h(A,B)   the Poincaré distance, or h-distance, between points A,B in H

Sets and relations in H which are analogs of sets and relations in H^2 are indicated by the use of "h" as an index or prefix to the corresponding notation in H^2. Thus $L_h(AB)$ is the h-line of points A,B in H, $\measuredangle_h BAC$ is the h-angle of h-rays $R_h[AB)$ and $R_h[AC)$, etc.

Chapter I. Some Historical Background

Section 1. The Parallel-Postulate Problem

Some understanding of basic arithmetic and geometric concepts, some knowledge of mathematical relations, were in all probability a part of human culture long before there was any recording of knowledge. The earliest records that do exist tend to confirm this view. Unfortunately, such records are not so much a story of culture at the time as they are clues to such a story. The interpretation of these clues by research scholars is constantly adding to our understanding of early civilizations, but the history of mathematical knowledge is far from complete.

If one asks not about the origins of mathematical knowledge but the origins of mathematical proof and the organization of mathematics as a subsystem of logic, there is fairly good agreement about an answer. It is not clear who first had the notion of proving a mathematical relation as a necessary logical implication of certain assumptions. The idea has been attributed both to Thales and Pythagoras. What does seem certain is that the notion of a mathematical proof crystalized as a concept and a method in Greek civilization somewhere around 600 B.C. And the period from 600 B.C. to 300 B.C. saw this "proof idea" exploited brilliantly by an extraordinary succession of gifted mathematicians. The brightest star of this galaxy, Archimedes, ranks with the best of all time, and Eudoxus, Appolonius and Euclid were also men of exceptional mathematical talent.

Somewhere around 300 B.C., the dates are not certain, Euclid wrote "Elements", the most famous of his works. In this treatise, he organized a large body of known mathematics, including discoveries of his own, into the first formal system of mathematics. This "formalness" was exhibited by the fact that "Elements" began with an explicit statement of assumptions called "axioms" or "postulates",

together with definitions. Other statements, called "theorems", were
then shown to follow necessarily from the axioms and definitions. The
work dealt primarily with mathematics which we now classify as geometry,
and the entire structure is what we now call "Euclidean Geometry".

Euclidean geometry was certainly conceived by its creators as
an idealization of physical geometry. The entities of the mathematical
system are concepts, suggested by, or abstracted from, physical experience, but differing from physical entities as an idea of an object differs from the object. However, a remarkable correlation existed between
the two systems. The angle sum of a mathematical triangle was stated
to be 180°, if one measured the angles of a physical triangle the
angle sum did indeed seem to be 180°, and so it went for a multitude
of other relations. Because of this agreement between theory and
practice, so to speak, it is not surprising that many writers came
to think of Euclid's axioms as "self evident truths". Centuries
later, the philosopher Immanuel Kant even took the position that
the human mind is essentially "Euclidean" and can only conceive of
space in Euclidean terms. Thus, almost from the its inception,
Euclidean geometry had something of the character of dogma.

Curiously, an early criticism of the "Elements", and one that
would ultimately lead to the discovery of non-euclidean geometry, had
little to do with the substance of the work. It concerned instead,
the following rather aesthetic question. If A_1, A_2, A_3, A_4, A_5, are five
statements that are axioms in a system, and if A_5 is logically implied
by the other four statements, then A_5 need not be assumed but can be
proved as a theorem, say theorem T. Whether A_5 is assumed or T is proved does not affect the system as a whole since the same body of implications can be derived from A_1, A_2, A_3, A_4, A_5 as from A_1, A_2, A_3, A_4, T. On
aesthetic grounds, however, it is obviously more satisfying to prove A_5
if it really is a theorem. What is most unsatisfactory of all is to
suspect that A_5 is really a theorem without being able to prove it. It
was exactly such a dissatisfaction that arose in connection with Euclid's

5th postulate or "parallel postulate". In present day language, this postulate asserts:

> if two lines r and s in a plane are inter-
> sected by a third line t in such a way that
> the interior angles on side of the transver-
> sal t have an angle sum less than 180° then
> r and s intersect on that side of the trans-
> versal t.

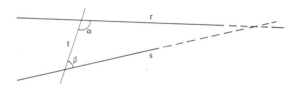

$$\alpha + \beta < 180° \Rightarrow r \cap s \neq \emptyset$$

The parallel-postulate drew attention to itself on two counts. First, unlike the earlier postulates it was a long and rather clumsy axiom. Second, Euclid derived a considerable body of theory before he made any use of the parallel-postulate. Whether for these or other reasons, the suspicion arose that the postulate was not independent of the others and could be proved. The problem of deriving it from the other axioms became known as the "parallel-postulate problem".

Section 2. The Lost Centuries

One of the oddest facts of history, and one that is not easy to explain, is that a span of roughly two thousand years separated the first formal system of mathematics, written by Euclid, from the criti-cal examination of that system in the 18th and 19th centuries which led to the discovery of non-euclidean geometry and our present under-standing of mathematical systems in general. A full explanation of this enormous gap in the development of geometry involves complex inter-relations of historical events. We will not attempt such an explanation, but will try to sketch an outline of this "interrupted story".

First, one must recognize that at Euclid's time, and for centuries before and after, formal education of any kind was restricted to a very small part of the population. And specialized "higher learning" was acquired by only a few in a still much smaller subgroup. Thus, to say for example, that such and such a discovery by Archimedes was "known in 250 B.C." probably means that perhaps a hundred or so scholars knew the facts of the discovery and that some of these had actually read and understood the manuscript. Greek geometry was an intellectual game, invented and developed by a few men of genius and studied by a small group of scholars. But there was no dissemination of this knowledge to the general public and virtually no application of it to technology.

During the period that spanned the rise and fall of the Roman Empire, roughly 100 B.C. to 500 A.D., the Romans must have had some awareness of the existence of Greek geometry, but they neither understood it nor valued it. And in Greek culture itself, the period of great mathematicial inventiveness was over. There were some exceptions. In the third century A.D., Pappus did original work in geometry and Diophantus made important contributions to number theory. But most of the mathematicial writing of these centuries consisted of scholarly commentaries on earlier works, and is valuable mainly as a record of those works, many of which have not survived. Over all, one can say of this period that mathematical knowledge, in any serious sense of the term, existed in the minds of a few Greek scholars and in the manuscripts at various centers of Greek culture, such as Constantinople and Alexandria.

After the fall of the Roman Empire, the Mohammedan conquest of

North Africa and parts of Spain, accomplished roughly between 600 and 900 A.D., put the sources of Greek mathematics in Arab hands. In general, they showed respect for culture and from the 7th to the 12th centuries they made extensive translations of the Greek manuscripts that came into their possession. Though Arab mathematicians knew Euclidean geometry and were aware of the parallel-postulate problem, their most important contributions were not in geometry but in algebra and trigonometry. But it was through an Arabic version of the "Elements" in a Moorish university that knowledge of Euclidean geometry found its way into western society.

In the 12th century Athelard translated the "Elements" from Arabic into Latin and this century also saw the founding of universities in England, France and Italy. Over the next few centuries, despite such deterrents as the plague and the Hundred Years War, knowledge of algebra, trigonometry and geometry diffused throughout Europe, a process that was greatly aided by the invention in the 15th century of the movable type printing press.

Probably the first European writer to discuss the parallel-postulate problem was Levi ben Gerson at the start of the 14th century. Knowledge of the supposed defect in Euclid's "Elements" became well established along with the geometry itself, and numerous attempts at a solution were made. These failed for the same reasons that a solution had eluded the Greeks. More will be said about this in the next section, where an account is given of a different attack on the problem made by Saccheri early in the 18th century.

Section 3. Saccheri and the "Near Miss"

At the start of the "Elements", Euclid gave an argument for what is commonly called the "side-angle-side" condition for the congruence of two triangles. Consider two triangles, $\triangle ABC$ and $\triangle DEF$, such that the sides $\overline{AB}$ and $\overline{DE}$ have the same length, the angles at A and at

D have the same size, and the sides $\overline{AC}$ and $\overline{DF}$ have the same length.

Euclid argued that $\overline{AB}$ could be superimposed on $\overline{DE}$ so that A and D

coincided, B and E coincided, and C was in the F-side of the line DE.

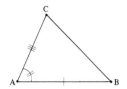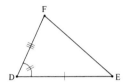

The congruence of the angles at A and D would then force the ray from

A through C to fall on that from D through F. Then, the congruence

of $\overline{AC}$ and $\overline{DF}$ would force C to coincide with F. The coincidence of

A,B,C with D,E,F respectively then implies that the angles at C and

F are congruent, those at B and E are congruent, and the sides $\overline{BC}$ and

$\overline{EF}$ are congruent.

Because Euclid had no definitions or axioms dealing with the

superposition of one figure on another, our present view is that he took

this congruence condition, in effect, as an axiom. He used this axiom

in establishing the following key theorem in the foundations of geome-

try, the so-called "weak exterior angle property".

Theorem A.

If ⦦DBC is an exterior angle to △ABC, then its measure is great-

er than that of the angle at A and that of the angle at C.*

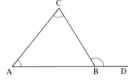

---

*Proposition 16, Euclid, T.L. Heath translation

The proof of this theorem went as follows. Let M be the midpoint of $\overline{CB}$ and on the line of A and M let E be such that M is the midpoint of $\overline{AE}$. In the triangles $\triangle AMC$ and $\triangle EMB$, we now have $\overline{AM}$ congruent to $\overline{EM}$,

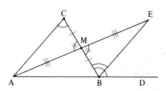

$\overline{MC}$ congruent to $\overline{MB}$ and $\angle AMC$ congruent to $\angle EMB$ because they are oppo-site or vertical angles. Thus, by the side-angle-side axiom, $\triangle AMC$ is congruent to $\triangle EMB$, and so the corresponding angles, $\angle MCA$ and $\angle MBE$, are con-gruent. Thus the angle at C is congruent to $\angle CBE$. But E is interior to the angle $\angle DBC$. Thus the measure of $\angle DBC$ is greater than that of $\angle CBE$, hence is greater than that of the angle at C. By an entirely similar argument, the measure of the exterior angle $\angle DBC$ is also great-er than that of the angle at A.

The heart of the proof just described lies in E's being interior to $\angle CBD$. For the proof to be valid, E has to exist on the line AM, twice as far from A as M is, and with M between A and E. For such a point E to exist, under all positions of A and M, it must be true that there is no upper bound to the distances between pairs of points on a line. That is, lines must be "unbounded in extent". This "infinite extent" of lines was freely used by Euclid , and by others for centuries afterwards, without the realization that the property constituted an unstated assumption. As we shall see presently, the oversight of this hidden axiom was to have odd consequences.

Using the weak exterior angle property, Euclid easily established the following theorem.

Theorem B

If a transversal t of two lines r and s is such that it forms with them a pair of congruent interior-alternate angles, then the

the lines r and s are parallel.*

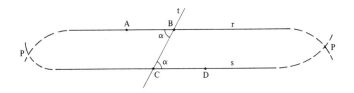

If the interior-alternate angles, ∢ABC and ∢DCB, have the same measure α, suppose that r and s intersect at point P. If P is in the A-side of t, then ∢DCB is exterior to ΔABP. By Th. A, the exterior angle ∢DCB is greater than ∢CBP, hence α > α, contradicting α = α. Similarly, if P is in the D-side of t, then ∢ABC is exterior to ΔBCP, and ∢ABC greater than ΔBCP again implies α > α, which is contradictory. Thus r and s cannot meet and so are parallel.

A natural question is whether or not the converse of Theorem B is valid. That is, if r and s are parallel (coplanar and non-intersecting), and if the transversal t forms with them interior-alternate angles ∢ABC and ∢DCB, must these angles be congruent? Euclid's answer to the question is regarded by many as a pure stroke of genius. What he said in effect is that the the angles are congruent if you assume the parallel postulate (c f. p.3)

To see the close connection between the parallel postulate and the converse of Theorem B, let r and s be parallel, let E on r be in the D-side of t, and let ∢ABC, ∢DCB, and ∢CBE have measures α, β , γ respectively. Suppose that α > β. Then, from α + γ = 180° and α > β, it follows that β + γ < 180°. Now the interior angles ∢DCB and ∢CBE, on

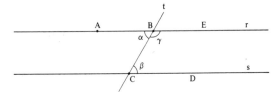

---

*Proposition 27, Euclid, T.L. Heath translation

the same side of t, have a sum less than 180° and so, by the parallel postulate, must intersect in the D-E-side of t, contradicting r parallel to s. Thus α > β is impossible. By an entirely symmetric argument α < β is impossible, and so α = β. Thus in Euclidean geometry, we have the following fact.

Theorem C

A transversal of two parallel lines forms with them congruent interior-alternate angles.

Euclid used the parallel postulate to prove Theorem C. But if he had taken Theorem C as an axiom, then he could have proved the parallel postulate. In this sense, Theorem C is equivalent to the parallel postulate. What the various attempts to prove the parallel postulate amounted to was simply the discovery of different properties equivalent to the parallel postulate. That is, in each of the proofs put forward somewhere an assumption was used that was just as basic as the parallel postulate itself. To list a few of these equivalents, if any one of the following theorems of Euclidean geometry is substituted for the parallel postulate, then the property of the postulate can be derived as a theorem.*

Theorem D

If point P is not on line t, then there is exactly one line through P that is parallel to t.

Theorem E

In a plane containing the line t, all points on one side of t and at a constant distance from t lie on a line.

Theorem F.

The sum of the measures of the angles in a triangle is 180°

---

*Theorems D, E, F are propositions 29, 30 and 32 in Euclid, T.L. Heath translation.

and in a quadrilateral is 360°

Theorem D as an axiom, is called the"Playfair Axiom", after the mathematician John Playfair (1748-1819), and is the parallel-postulate equivalent commonly used in present day textbooks. The property in Theorem E has a long history in false proofs for the parallel postulate clear back to the time of Euclid. It was the property in Theorem F, however, that formed the keystone in a work written by Gerolamo Saccheri in 1773. This work nearly resolved the parallel-postulate problem and is one of the great "near-misses" of history.

Without assuming the parallel postulate, Saccheri considered a quadrilateral ABCD, with right angles at A and B and with congruent "sides" $\overline{AD}$ and $\overline{BC}$. Segment $\overline{AB}$ is the "base" of such a quadrilateral,

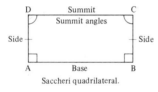

Saccheri quadrilateral.

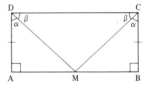

$\overline{CD}$ is the "summit", and the angles at C and D are the "summit angles". If M is the midpoint of the base, then ΔMBC is congruent to ΔMAD, by side-angle-side, so ∢MCB and ∢MDA have a common measure α. Since $\overline{MC}$ is congruent to $\overline{MD}$, ΔMCD is isosceles and its base angles ∢MCD and ∢MDC have a common measure β. Thus the summit angles at C and D have the common measure α + β , so these angles are congruent. Saccheri saw that if he could prove that the summit angles were necessarily right angles, he would have a special class of quadrilaterals with a 360° angle sum. From this he could show that all quadrialterals have such an angle sum and that all triangles have a 180° angle sum. With this last fact, it is easy to prove the parallel-postulate.

Saccheri's plan to establish that the summit angles must be

right angles was itself an ingenious new attack on the problem. He would first assume that the summit angles were obtuse and show that this led to a contradiction. Next, he would assume that the summit angles were acute and show that this was also contradictory. With right angles left as the only possibility, he would have the key fact that he needed.

Starting with the obtuse angle hypothesis, Saccheri did reach a contradiction. But in doing so, he used the weak exterior angle property and hence the hidden assumption that lines are of unbounded extent. And it was this unstated axiom that the obtuse angle hypothesis was contradicting. Thus what Saccheri proved, without knowing it, was that if lines are of unbounded extent then the angle sum of quadrilaterals is equal to or less than 360°.

When Saccheri attempted to eliminate the possibility of acute summit angles, he found the task much more difficult. He may, in fact have felt some dissatisfaction with his reasoning since he gave more than one argument for the supposed contradiction. However, none of these was correct. The proofs involved mistakes with limit processes and an unjustified use of "points at infinity".

We know now that the acute summit angle hypothesis, taken as a substitute for the parallel postulate, does not lead to any simple contradiction. Instead, in combination with Euclid's other axioms, it forms a basis for a system, now called "Hyperbolic geometry", which is different from Euclidean geometry, but whose structure is no less consistent. Many of the properties that Saccheri derived, before he reached what he supposed was a contradiction, are now standard theorems in this new geometry.

Saccheri's work was a major achievement, and a landmark in the long history of attacks on the parallel-postulate problem. The work is also a paradox. Saccheri entitled it "Euclid Vindicated of All Flaw". But since Euclid had taken the property of the parallel postulate

as an axiom, his judgement in doing so would be vindicated if one
could show that this property could not be derived from the other
axioms. In Saccheri's approach, where in effect the parallel postulate
is denied, a vindication of Euclid would be to show that such a denial
does not imply a contradiction. But it is much harder to show that a
system cannot imply a contradiction than it is to obtain a contradiction
from an inconsistent set of axioms. Thus Saccheri was committed to an
almost hopelessly difficult task, unless Euclid was wrong. Euclid had,
in fact, made a mistake in not stating the "unbounded extent of lines"
axiom. This axiom is inconsistent with the obtuse summit angle hypothesis, and Saccheri established the inconsistency, though he mistook its
source. However, Euclid had not made a mistake about the independence
of the parallel postulate. Neither it, nor the acute summit angle
hypothesis, is inconsistent with the other axioms, including the unbounded extent of lines. Thus in this instance we have the odd situation
of Saccheri attempting to vindicate Euclid, for the wrong reasons, and
failing to genuinely vindicate Euclid only because of mistakes that he
himself made. Except for those mistakes, Saccheri might well have
been the discoverer of the first non-euclidean geometry.

Section 4. The Correct Perspective—Gauss, Bolyai, Lobatchevsky

A clear and unequivocal development of the hyperbolic geometry
that Saccheri just missed did not occur until nearly a century later.
Two facts are helpful in understanding this delay. First, Saccheri,
like others before him, saw the parallel-postulate problem as an
aesthetic one. As the title of his work indicates, he was removing
a blemish from Euclid's axiom system. Thus Saccheri's reasoning,
whether or not it was correct, did not imply any radical consequences.
His work caused some initial excitement and then became a "minor paper"
known only to a few specialists.

A second fact that helps to explain the neglect of Saccheri's work is the state of mathematics at the time. In 1629, roughly a century before Saccheri, Descartes had brought algebra and geometry into a new fusion with his method of coordinate geometry. This, in turn, was a precursor of the Calculus discovered by Newton and Leibniz toward the end of of the 17th century. Mathematicians of the 18th century were involved with the ramifications of these new methods and with the development of whole new areas of mathematics. Classical Euclidean geometry, including the parallel-postulate question, was relegated to the wings while these new developments held center stage.

The most notable exception to the general neglect of the parallel postulate problem, after Saccheri, was a paper by J. Lambert, "The Theory of Parallel Lines", published posthumously in 1786. In this paper, Lambert considered quadrilaterals with three right angles - now called "Lambert Quadrilaterals" - and discussed some consequences of the fourth angle being acute. Lambert probably knew Saccheri's work, and in his own paper he made some prophetic speculations that were an advance over Saccheri's results. But it is doubtful that Lambert, any more than Saccheri, saw the full significance of his own suggestions.

After a revolutionary idea is clearly formulated by someone, it is a common sport of historians to trace out all the predecessors whose work contains some suggestion of the idea. Since the judgement of "how close" they came is usually subjective, there is plenty of room for speculation. However, if one disregards this "hindsight guessing", it seems clear that the first person to see the parallel-postulate problem from the correct perspective was the famous mathematician Carl Friedrich Gauss (1777-1855). Not only did he see that Euclid was right, and that the parallel postulate was independent of the other axioms, he also saw that this implied the existence of a geometry different from that of Euclid.

It seems that Gauss was still a schoolboy when he reached the conviction that the parallel-postulate cannot be proved. Throughout his long lifetime, which was prolific in mathematical and scientific discoveries, Gauss returned time and again to the development of the geometry that is implied by Saccheri's acute summit angle hypothesis. But we know this only from his correspondence and from his private papers which became available after his death. He never published his results and this omission prevented him from being credited, officially, with the discovery of hyperbolic geometry.

Two contemporaries of Gauss not only worked out the properties of hyperbolic geometry, as he had done, but also published their discoveries. These two, John Bolyai (1802-1860) and Nicolai Lobatchevsky (1792-1856) worked independently of each other and both are now credited with the discovery of non-euclidean geometry. The first published development of the new geometry occurred in 1829 in an article written by Lobatchevsky for the "Kazan Messenger". This paper, in Russian and in a rather obscure journal, went unnoticed by the scientific world. In 1832, John Bolyai's discoveries were published in an appendix to a work called "Tentamen", written by his father, Wolfgang Bolyai. Gauss, who was a lifelong friend of the father, received a copy of the Appendix in 1832 and instantly recognized it as the work of a genius. By then Gauss was famous, and praise from him could have made John Bolyai's reputation. But apparently Gauss' determination to avoid all disputes about non-eucldean geometry not only included the suppression of his own work, but silence about the work of others. He made no public statements about the new geometry, and it was years later before the nature of Bolyai's achievement was understood.

In 1837, Lobatchevsky published a paper in Crelle's journal about the new geometry, and in 1840 he wrote a small book in German about the same subject. Again Gauss recognized genius and in 1842 he proposed Lobatchevsky for membership in the Royal Society of Gøttinger

But Gauss made no public statements about Lobatchevsky's work, nor did
he tell Lobatchevsky about John Bolyai. It seems that the latter
learned about Lobatchevsky somewhere around 1848, but that Lobatchevsky
died without ever knowing that his geometry had a co-discoverer.

Just before his death, and when he was blind, Lobatchevsky dic-
tated a detailed account of his revolutionary ideas about geometry.
This work entitled "Pangeometry" was translated into French by Jules
Houel in 1866. In the following year he also translated Bolyai's
work into French. These two translations inspired others, as well
as a host of expository articles, and knowledge of the new geometry
began to be widespread. The parallel-postulate problem was finally
put to rest, Euclid really was vindicated, and the way in which the
problem was solved changed forever man's understanding of mathematical
systems.

Section 5. Not-Euclidean Geometry, Absolute Geometry

It is our intention in this book to derive the basic structure
of Hyperbolic plane geometry, using the same elementary methods as
did Bolyai and Lobatchevsky. In the previous sections we have sketched
in outline the long history behind the discovery of this new geometry.
The discovery itself brought on such a flood of new information about
the nature of geometric systems that even a summary of these developments is beyond the scope of this work. However, we do want to comment
on some particular items that are either directly related to the methods
we will use or that serve to put the present work in a broader perspective.

One rather natural outcome of the discoveries made by Bolyai
and Lobatchevsky was a critical reppraisal of Euclidean geometry itself.
The modernization of Euclidean geometry, together with an analysis of
its axioms, was the work of such men as M. Pasch (1843-1930), G. Peano

(1858-1932) and D. Hilbert (1862-1943). In this analysis, the role
played by the "unbounded extent of lines" came to be understood. It
was discovered that with a suitable modification of this axiom,
Saccheri's obtuse summit angle hypothesis also led to a new geometry.
Thus, there was a geometry for each of Saccheri's three hypotheses.
It was F. Klein (1849-1925), borrowing an idea from projective geometry, who suggested that the geometries corresponding to the obtuse,
the right, and the acute summit angles be called "elliptic", "parabolic"
and "hyperbolic" respectively. The names have become standard, though,
of course, in most contexts parabolic geometry is referred to as "euclidean".

As one might guess, once the notion of "a geometry" was established as simply a mathematical system whose basic elements had properties analogous to the "points", "lines" and "planes" of classical
geometry, then many new geometries were devised. The term "non-euclidean" came to refer to either elliptic or hyperbolic geometry, and
"not-euclidean" was used for geometries that were not elliptic, hyperbolic, or euclidean. An especially important development was the discovery that many of the new geometries had euclidean representations.
This fact was used by such men as E. Beltrami (1835-1900), A. Cayley
(1821-1859) and F. Klein to show that both elliptic and hyperbolic
geometry were no less consistent than euclidean geometry. We will
expand on this topic in our concluding chapter.

The critical re-analysis of Euclid's work had one consequence
we need to describe because it relates directly to an axiom system
we will use. From the very beginnings of formal geometry, the metric
properties of the euclidean line were inextricably related to the
properties of real numbers. Even before Euclid, it was discovered
that two segments could be such that no segment was an integral divi-
sor of both. In other words, the lengths of the two segments could
be incommensurable or, equivalently, the ratio of their lengths could

be an irrational number.

Since neither a proper theory of limits, nor a systematic
foundation for the real numbers, existed at the time of Euclid, his
handling of certain topics, for example, the proportions in similar
figures, was unsatisfactory by modern standards. The difficulties,
however, were easier to detect than to solve. Unless one assumed the
properties of real numbers (the familiar "number line" of elementary
arithmetic) it was necessary to employ some continuity axiom and to
derive real number properties in geometric form. Such a program, for
example, was carried out in Hilbert's development of Euclidean Geometry,
and it revealed the genuine sophistication of this familiar and intuitive subject. However, this very sophistication, though satisfying
to the "purist", made the Hilbert axioms unsuitable for elementary
studies of Euclidean geometry. In 1932, George Birkhoff proposed
a far less stringent set of axioms that permit a quick and accurate
derivation of Euclidean geometry.

This is the axiom system we will use, and the use will involve
still another geometry that needs some explanation.

In Section I, we mentioned that one factor which drew attention
to the parallel postulate was the amount of theory that Euclid developed before he made any use of the postulate. When the idea of "different
geometries" became commonplace, it was recognized that that part of
Euclidean geometry which does not depend on the parallel postulate
(or an equivalent) is itself an important geometry. All the theorems
of this system, commonly called " bsolute Geometry", are valid in
both Euclidean and Hyperbolic geometry. It is precisely because one
has the familiar theorems of Absolute Geometry to build on that the
derivation of Hyperbolic geometry is relatively simple. However,
though the relations of Absolute geometry are familiar from Euclidean
geometry, the precise extent of Absolute geometry, within the larger
system, is not ordinarily identified in beginning courses. It is this

identification that we will review in the next chapter, based on the Birkhoff axioms.

　　To conclude this historical sketch, some mention should be made of the physical implications of not-euclidean geometry. As we said at the outset, Euclidean geometry had the character of dogma because it was thought to be the necessary mathematical representation of the properties of physical space. The discovery of Hyperbolic geometry did not, in itself, invalidate that opinion. But as different geometries became familiar, and as revolutionary advances were made in Physics and Astronomy, the Euclidean-Newtonian framework for physical space came more and more to be viewed as an approximation of a much more complex reality. It is not an easy matter to explain what "best" means in the question "What is the best geometric representation of physical space?", and there is, at present, no certain answer to the question. We will not attempt to deal with the problem in this book.

Chapter II. Absolute Plane Geometry

Introduction

In this chapter we will review the basic properties of Absolute Plane geometry, based on the Birkhoff axioms. All the theorems to be considered are also theorems of Euclidean geometry and hence, for the most part, will be familiar to the reader. For this reason, proofs will not be given except for a few theorems not ordinarily encountered in a beginning course in Euclidean geometry. Our intent is to sketch a natural progression for the theorems and to introduce notations and conventions that we will need in the later study of Hyperbolic geometry.

Though this chapter can be read quickly, and serves mainly for reference, its importance can scarcely be overstated. The hyperbolic axiom, to be introduced in the next chapter, seems to violate common sense. Its implications are therefore not a matter of intuition but of logic. To deduce those implications carefully, it is indispensible to know precisely the body of established facts from which the logic proceeds.

Section 1. Linear Sets and Linear Order

We suppose that we are dealing with a set, called the "Absolute Plane", whose elements are points. We will denote points by capital English letters, such as A,B etc.* The absolute plane is a universal set, or space, in the sense that all points belong to the plane. We

---

* Because we feel that a rigid insistence on the distinction between use and mention causes more confusion than it prevents, we shall sometimes treat a formula or symbol as its own name, and not place it in inverted commas when it is referred to.

will use the italic form of capital English letters, such as R, S, T, as variables for subsets of the plane and will denote the plane itself by A^2 (where the exponent suggests dimension). If S is any subset of A^2, the complement of S denoted by $Cp(S)$, is the set of points not in S. The complement of A^2 is the empty set, or null set, $\emptyset$ that has no elements.

Axiom 1.

      There exist non-empty subsets of the plane called "lines", with the property that each two points belong to exactly one line.

      It will be useful to have the agreement that when the number of objects in some collection is given by a specific number, such as "two" or "five" we mean precisely that number of distinct objects and not that number of symbols for objects some of which may denote the same object. Thus, "two points" in axiom 1 automatically has the meaning of "two distinct points". On the other hand, should we refer to "points A and B", without specifying any number, then we do not exclude the possibility that A and B denote the same point.

Line Conventions

      Lines will commonly be designated by lower case English letters such as r, s, t, etc. If A and B are two points, we will also use either L(AB) or L(BA) to represent the unique line to which they belong. A set is a linear set if it is contained in some line, otherwise it is a non-linear set. The points of a linear set are collinear and those of a non-linear set are non-collinear. Two or more sets are collinear or non-collinear according to whether their union is a linear or non-linear set. Lines r and s are intersecting or non-intersecting according as r ∩ s ≠ ∅ or r ∩ s = ∅. If r and s are distinct lines and if A ∈ r ∩ s, then r and s intersect at A.

Theorem 1.

If two lines intersect, they intersect at only one point.

The next axiom is suggested by the physical experience of measuring distance between points.   If we always keep to the same unit, whether it is inches or yards or centimeters, then we expect the distance of
two points to be a unique number.   Axiom 2 amounts to regarding distance
as given by some fixed unit of length.

Axiom 2

Corresponding to points A, B there exists a unique non-negative
number $d(A,B) = d(B,A)$ which is the distance between A and B which is
zero if and only if $A = B$.

If a ruler is laid alongside a physical line then the numerical
marks on the ruler are next to points on the line.   If point A on the
line is next to the mark 3 on the ruler we know that A is 3 units away
from the point of the line next to the zero end of the ruler.   This
idea of assigning numbers to points and using the numbers to determine
the distance between the points is abstracted and generalized in the
following axiom that is often called the "Birkhoff ruler axiom".

Axiom 3

If t is a line and R is the set of all real numbers, there exists
a one to one correspondence, denoted by X <-> x, between the points X
on t and the numbers x in R such that the distance between points A,B
on t is the absolute value of the difference of the numbers a, b in R
which correspond to A and B respectively.

Since there are infinitely many real numbers Axiom 3 implies that
every line has infinitely many points and this with Axiom 1, implies
that there are infinitely many points in A^2.   Furthermore, Axiom 3 implies that there is no upper bound to the distances between pairs of
points on a line.   Thus Axiom 3 implies that lines in A^2 have the "un-

bounded extent" property mentioned in I-3. Axiom 3 does not hold in Elliptic Geometry (see I-5), where lines are limited in extent.

In the correspondence X <-> x of Axiom 3, the number x is said to be a <u>coordinate</u> of X and the correspondence is a system of <u>metric co-ordinates</u> for the line. In measuring physical distances along a line with a ruler, one can obviously slide the ruler to a new starting point, or turn the ruler around and measure distances in the opposite direction The next theorem gives a mathematical analog of these physical processes

<u>Theorem 2.</u>

If X <-> x is a system of metric coordinates for line and if h is a fixed number, then the correspondence X <-> x + h is also a system of metric coordinates for the line. Also, the correspondence X <-> - x is a metric coordinatization of the line.

<u>Theorem 3.</u>

If A and B are two points of line t, there exists a system of metric coordinates for t, X <-> x, such that A <-> o, B <-> b, and b > o.

Metric coordinates and the natural order of the real numbers can now be used to establish order for points of a line, and this order, in turn, provides a way of defining the basic linear sets namely segments and rays.

<u>Definition</u>

(Betweeness of numbers)　A real number b is <u>between</u> two numbers a and c if the magnitude of b is intermediate to those of a and c. That b is between a and c will be denoted by either <abc> or <cba> and either implies that the numbers a,b,c satisfy one of the relations a < b < or c < b < a.*

---

*　　A more logical practice would be to define <abc> to mean that b
　　is between a and c and then to show that <abc> implies and is im-

Definition

(Betweeness of points)   A point B is <u>between</u> points A and C, denoted by either <ABC> or <CBA>, if A,B,C are three collinear points
and $d(A,B) + d(B,C) = d(A,C)$.

Theorem 4

If A,B,C are three points on line t, then exactly one of them is
between the other two, and B is between A and C if and only if in every
metric coordinate system for t the coordinate of B is between the coordinates of A and C.   That is, <ABC> if and only if <abc>.

Definition

(Segments) If $A \neq B$, the <u>open segment</u> of A and B, denoted by either
S(AB) or S(BA), is the set defined by
$$S(AB) = S(BA) = \{X: <AXB>\}$$
The <u>closed segment</u> of A and B, or simply the segment A and B, is the
set denoted by either S[AB] or S[BA] and defined by
$$S[AB] = S[BA] = \{X: X=A, \text{ or } <AXB>, \text{ or } X=B\}.$$
The <u>half-open</u> segments of A and B are the sets defined by
$$S[AB) = \{X: \quad X=A, \text{ or } <AXB>\}$$
and
$$S(AB] = \{X: <AXB>, \text{ or } X=B\}.$$
The points A,B are the <u>endpoints to</u> all four segments, and $d(A,B)$ is the
length of S[AB].

Definition

(Rays)   If $A \neq B$, the <u>open ray from</u> A <u>through</u> B is the set

plied by <cba>. Our shortcut is simply a way of saying that this proof
is obvious.

denoted by R(AB) and defined by

$$R(AB) = \{X: <AXB>, \text{ or } X=B, \text{ or } <ABX>\}.$$

The closed ray from A through B is the set denoted by R[AB) and defined by

$$R[AB) = \{X: X=A, \text{ or } <AXB>, \text{ or } X=B, \text{ or } <ABX>\}.$$

Point A is the origin to both the open and closed rays.*

Comment

It is important to observe that in the notations S[AB), S(AB],
R(AB) and R[AB) the pair "AB" has the character of an ordered pair,
and changing the order changes the set. For example, S[AB) $\neq$ S[BA)
and R(AB) $\neq$ R(BA).

Definition

(Opposite rays)   Two rays are opposite rays if they are collinear,
have a common origin, and their intersection is empty or else consists
only of their common origin.

Theorem 5

If X <-> x is a coordinatization of line t, and A on t has coordinate
a, then there are exactly two open rays on t with origin A. They are
opposite rays and are the sets

$$R: = \{X: x > a\}$$

and $$R': = \{X: x < a\}.$$

There are also exactly two closed rays on t with origin A. They are
opposite rays and are the sets

$$S = \{X: x \overset{\geq}{} a\}$$

and $$S' = \{X: x \overset{\leq}{} a\}$$

---

*   Open and closed rays are also called open and closed "half-lines".

Corollary 5.1

A ray is uniquely determined by its origin and any one of its other points. That is, C ∈ R(AB) implies that R(AC) = R(AB).

Corollary 5.2

Point B is between points A and C if and only if R(BA) and R(BC) are opposite rays.

Corollary 5.3

If A ≠ B, R(AB) ∩ R(BA) = S(AB) and R[AB)∩ R[BA) = S[AB].

Theorem 6

If h is a positive number, there is exactly one point C on R(AB) whose distance from A is h, and <ACB>, C = B, or <ABC> according as
$$d(A,C) < d(A,B), \quad d(A,C) = d(A,B), \quad \text{or} \quad d(A,C) > d(A,B).$$

Theorem 7

If X <-> x is a coordinatization of L(AB), with A <-> a and B <-> b, then S[AB] = {X: x=a, or <axb>, or x=b}, S(AB) = {X: <axb>}, S[AB) = {X: x=a, or <axb>}, S(AB] = {X: <axb>, or x=b}.

Corollary 7.1

Point C on L(AB) is between A and B if and only if d(C,A) <d(A,B) and d(C,B) <d(A,B).

Corollary 7.2

If C and D are points of S(AB), then d(C,D) <d(A,B).

Theorem 8

If A is a point of line t and h is a positive number, there are exactly two points of line t at distance h from A and A is between them.

Theorem 9

On the line L(AB) there is exactly one point that is equidistant from A and B and this point is between A and B.

The point in Th. 9 is called the "midpoint" of S[AB] and is said to "bisect" the segment.

Definition

(Like and opposite directed rays) A ray R and a ray S are like directed if they are collinear and one contains the other. They are opposite directed if they are collinear and neither contains the other. Clearly two collinear rays are either like or opposite directed and not both.

Theorem 10

Ray R is like directed to ray S if and only if the open and closed rays of R are like directed to S and if and only if R is like directed to the open and closed rays of S.

Theorem 11

Every ray is like directed to itself and opposite rays are opposite directed.

Theorem 12

Like directedness of collinear rays is transitive, that is if rays R and S are like directed and if rays S and T are like directed, then rays R and T are like directed.

We conclude this section with a concept that will play an important role throughout our entire study.

Definition

(Convex set)   A set s is <u>convex</u> if A $\in s$ and B $\in s$ and $<AXB>$ imply that X $\in s$.

Theorem 13

The intersection of any number of convex sets is again a convex set.

Theorem 14

The plane A^2, the empty set, singleton sets, lines, rays and segments are convex sets.

Section 2. Half-Planes, Angles, and Angle Measure

Using the order of the real numbers, it has been established that if A is a point of line t, then t is the union of {A} with two opposite open rays on t with origin A, say R(AB) and R(AC). The three sets, {A} R(AB), and R(AC), are non-empty and convex, are pairwise disjoint (no two of them intersect), and if X $\in$ R(AB) and Y $\in$ R(AC) then A $\in$ S(XY). We now introduce the Birkhoff "plane separation axiom" to establish that each line gives a similar division of the plane into opposite halves.

Axiom 4

Corresponding to a line t, there exist exactly two non-empty, convex sets R' and R'' with the following properties: (i) the plane is the union of the sets R', R'' and t; (ii) no two of these three sets intersect; (iii) X $\in R'$ and Y $\in R''$ imply that S(XY) intersects t.

Definition

(Half-Planes)   Corresponding to a line t, the unique sets R'

and R'' in Axiom 4 are the <u>opposite</u> <u>sides</u> or <u>opposite</u> <u>open</u> <u>half-planes</u> of t. The sets R' ∪ t and R'' ∪ t are the <u>opposite</u> <u>closed</u> <u>half-planes</u> of t. Line t is the <u>edge</u> to all four half-planes.

Half-plane Conventions

If R' and R'' are the opposite sides of line t and if A is not on t, then from properties (i) and (ii) of Axiom 4, A belongs to exactly one side of t. Thus if A ∈ R' we will call R' the "A-side of t" and use the notation H(t;A) to denote R'. Similarly, if B ∈ R'', then H(t;B) will also denote R''. The opposite closed half planes will be denoted by H[t;A) and H[t;B).
The following concept will prove to be useful.

Definition

(Separation of sets by a line)   Sets R and S are <u>separated</u> by line t if they are non-empty sets and one side of t contains R and the other contains S. <u>Points</u> A and B are <u>separated</u> by t if A and B lie in opposite sides of t.

Theorem 1

Two points A,B lie in one side of line t if and only if S[AB]∩ t = ∅

In the proofs for order relations in the plane, the following "edge-ray" theorem plays a crucial role and we will appeal to it frequently in later work.

Theorem 2

If point A is on line t and point B is not, then every point of ray R(AB) is in the B-side of t, and line t separates R(AB) and the open ray opposite to R(AB).

Corollary 2

Closed half-planes are convex sets.

Definition

(Angle and angle interior)   An <u>angle</u> is the union of two closed,
non-collinear rays that have a common origin. The closed rays are the
<u>arms</u> of the angle and the common origin is the <u>vertex</u> of the angle.
If rays R[BA) and R[BC) are non-collinear, the angle that is their
union is denoted by either ⟩ABC or ⟩CBA. The <u>interior</u> of the angle
is the set   In(⟩ABC) = A-side of L(BC) ∩ C-side of L(BA).

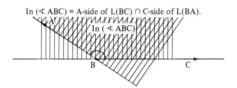

In (∢ ABC) = A-side of L(BC) ∩ C-side of L(BA).

Theorem 3

Angle interiors are convex sets.

Definition

(Opposite angles, angles of intersection of two lines)
Two angles are <u>opposite</u> <u>angles</u> if the arms of one are the closed rays
opposite to the arms of the other. If R[BA) and R[BA') are opposite
rays, and if R[BC) and R[BC') are opposite rays, and A,B,C are non-collinear, then ⟩ABC and ⟩A'BC' are opposite angles, and ⟩ABC' and ⟩A'BC are
opposite angles. The four angles are the <u>angles</u> <u>of</u> <u>intersection</u> of
lines L(BA) and L(BC).*

Definition

(Adjacent angles)   Two angles are <u>adjacent</u> if they have an arm

---

*

Opposite angles are also called "vertical angles".

in common and their interiors do not intersect.

We now make use of angles and angle interiors to introduce the notion of betweeness for rays.

<u>Definition</u>

(Betweeness of rays)   Ray s is <u>between</u> rays R and T if: (i) R , S and T have a common origin;   (ii) the closed rays of R and T form an angle; (iii) the open ray of s is interior to the angle of R and T.

<u>Theorem 4</u>

Ray s is between rays R and T if and only if both the open and closed rays of s are between the following pairs of rays: the open rays of R and T ; the closed rays of R and T ; the open ray of R and the closed ray of T; the closed ray of R and the open ray of T.

<u>Theorem 5</u>

Point P is interior to ⧧ABC if and only if R(BP) is between R(BA) and R(BC).

<u>Theorem 6</u>

If A,B,C are non-collinear, then S(AC) ⊂ In(⧧ABC).

<u>Theorem 7</u>

If P ∈ In(⧧ABC), then R(BP) intersects S(AC) and L(BP) separates R(BA) and R(BC). The interiors of ⧧ABP and ⧧CBP are disjoint and each is a proper subset of In(⧧ABC).

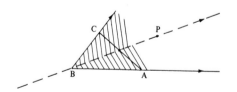

Theorem 8

If ray R(BP) is interior to ⊀ABC then the opposite open ray is interior to the angle opposite to ⊀ABC.

Theorem 9

If <ABC> on line t, and if points D and E in one side of t are
non-collinear with B, then exactly one of the rays R(BD) and R(BE) is
between the other and ray R(BC). If R(BE) is between R(BD) and R(BC)
then R(BD) is between R(BE) and R(BA).

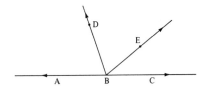

If physical angles are measured with a protractor, marked in degrees,
then each angle has a degree measure between zero and 180. We now introduce the Birkhoff axioms that give an analogous measurement for mathematical angles.

Axiom 5

To each angle ⊀ABC there corresponds a unique real number between
0 and 180, denoted by ⊀ABC°, which is the <u>degree measure</u> or simply <u>the
measure</u> of the angle.

If the degree measure of an angle is x, we call the angle "an angle
of x degrees" and sometimes write "x°" to indicate that x is an angle
measure.

The next two axioms are called the "angle addition" and "protractor"
axioms respectively.

Axiom 6

If R(BD) ⊂ In(⊀ABC), then ⊀ABD° + ⊀DBC° = ⊀ABC°.

Axiom 7

If R[AB) is a ray in the edge t of an open half plane H(t;P),
then there exists a one to one correspondence between the open rays
in H(t;P)which have origin A and the set of real numbers between 0 and
180 such that if R(AX) and x correspond then ⋊BAX⁰ = x.

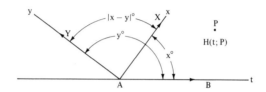

The correspondence R(AX) <-> x in Axiom 7 will be called "a system of
ray coordinates for the open half-plane H(t;P) with respect to the
base ray R[AB)", and x will be called "a coordinate" of R(AX).

Theorem 10

If R(AX) and R(AY) are two rays in a side of line L(AB) that have
coordinates x and y respectively with respect to a base ray R[AB),
then ⋊XAY⁰ = |x - y|. That one of the rays R(AX) and R(AY) with the
smaller coordinate is between the other and R[AB).

Corollary 10

If X and Y lie in one side of line L(AB) then R(AX) = R(AY) if
and only if ⋊XAB⁰ = ⋊YAB⁰, and R(AY) is between R(AX) and R(AB) if and
only if ⋊YAB⁰ < ⋊XAB⁰.

Definition

(Angle classifications) An angle is an _acute_ angle, is a _right_
angle, or is an _obtuse_ angle according as its measure is less than,
is equal to, or is greater than 90⁰. Two angles are _complements_ if the
sum of their measures is 90⁰ and are _supplements_ if the sum of their

measures is 180°.

Theorem 11

    If R(BA) and R(BC) are opposite rays that are not collinear with
R(BD) then ⅋DBA and ⅋DBC are supplements.

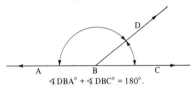

Theorem 12

    If ⅋DBA and ⅋DBC are adjacent and supplements then R(BA) and R(BC)
are opposite rays.

Theorem 13

    Opposite angles have the same measure.

The following theorem is sometimes useful in establishing collinearity.

Theorem 14

    If <ABC> on line t, and if t separates D and E, then B is between
D and E if and only if ⅋ABD° = ⅋CBE°.

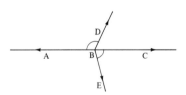

Definition

    (Angle bisectors)   The open and closed rays R(BD) and R[BD)
bisect ⅋ABC if R(BD) ⊂ In(⅋ABC and ⅋DAB° = ⅋DAC°. A line bisects an
angle if it contains a ray that bisects the angle.

Theorem 15

    There is exactly one open ray that bisects an angle and its line is the unique line that bisects the angle.

Theorem 16

If a ray bisects an angle, its opposite (open and closed) rays
bisect the opposite angle.

Definition

(Perpendicular lines)   Two lines are perpendicular if they intersect and one of their angles of intersection is a right angle.   That
lines s and t are perpendicular will be denoted by s ⊥ t or t ⊥ s.   If
A is their intersection point, the lines are perpendicular at A.

Theorem 17

If one of the angles of intersection of two lines is a right angle
then all four are right angles.

Theorem 18

If A is a point of line s then there is exactly one line that is
perpendicular to s at A.

Section 3. Triangle Relations, Congruence, Foot in a Set

In this section, we review some familiar conventions about a basic
figure of plane geometry, namely the triangle.

Definition

(Triangle, triangle interior)   If A,B,C are three non-collinear
points, the union of the closed segments, S[AB] ∪ S[BC] ∪ S[CA], is the
triangle with vertices A,B,C. The union is denoted by the symbol "Δ"
followed by the letters A,B,C in any order, for example "ΔBCA".   The
closed segments are the sides of the triangle, and ∢ABC, ∢BCA, ∢CAB
are the angles of the triangle. An angle and its vertex are both said
to be opposite to the side to which the vertex does not belong.   The
interior of the triangle, denoted by In(ΔABC) is the set

In(ΔABC) = A-side of L(BC) ∩ B-side of L(CA) ∩ C-side of L(AB).

Definition

(Exterior angle to a triangle)   An angle that is adjacent to an angle of a triangle and supplementary to it is an exterior angle to the triangle.

Theorem 1

The interior of a triangle is a convex set.
The following relations concerning the intersection of a line and a triangle are sometimes referred to as "Pasch properties" of a triangle.

Theorem 2

A line that intersects a triangle intersects at least two sides of the triangle. A line through a point that is interior to a triangle intersects the triangle at exactly two points. If one of these intersections is a vertex, then the other belongs to the open segment of the opposite side.

Definition

(Congruent segments and angles)   Closed segments are congruent if they have the same length. Angles are congruent if they have the same measure. The symbol "$\cong$" is used for the congruence relation. Thus, S[AB] $\cong$ S[CD] if d(A,B) = d(C,D)   and $\angle$ABC $\cong$ $\angle$DEF if $\angle$ABC^0 = $\angle$DEF^0

Definition

(Correspondence of triangles)   A correspondence of $\triangle$ABC with $\triangle$DEF is a one-to-one correspondence of the vertices, a one-to-one correspondence of the sides, and a one-to-one correspondence of the angles such that the vertices of corresponding angles are corresponding vertices of the triangles, and sides that correspond are opposite to vertices that correspond. The notation ABC <-> DEF will be used to denote the correspondence in which A <-> D, B <-> E, C <-> B, $\angle$A <-> $\angle$D, $\angle$B <-> $\angle$E $\angle$C <-> $\angle$F, S[AB] <-> S[DE], S[BC] <-> S[EF], S[CA] <-> S[FD]. Two correspondences are the same if the same triangle parts are paired in both.

Thus ABC <-> DEF and BCA <-> EFD are the same correspondence.

Definition

(Congruence of triangles)   A correspondence of triangles is a congruence of the triangles if the corresponding sides are congruent and the corresponding angles are congruent. The triangles ΔABC and ΔDEF are congruent triangles if at least one of their correspondences is a congruence. The notation ΔABC ≅ ΔDEF will be used to state that the correspondence ABC <-> DEF is a congruence of the triangles.

To state the triangle congruence conditions in familiar language, an angle of a triangle is said to be "between" the two sides it contains, and a side is said to be "between" the two angles whose vertices are its endpoints. We now take the "side-angle-side" or "s- x-s" condition as an axiom.

Axiom 8

If a correspondence of two triangles, or of a triangle with itself, is such that two sides and the angle between them are respectively congruent to the corresponding two sides and the angle between them, the correspondence is a congruence of the triangles.

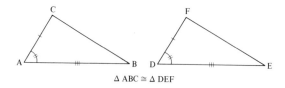

$\Delta\ ABC \cong \Delta\ DEF$

With Axiom 8, one can establish the "weak exterior angle" property that plays a crucial role in both Absolute and Hyperbolic geometry (c.f. Th. A, I-3).

Theorem 3

An exterior angle to a triangle is greater than either of the non-adjacent angles of the triangle.

Corollary 3

If one angle of a triangle is not acute then the other two angles
of the triangle are acute.

Definition

(Classification of triangles)   A triangle is an obtuse triangle,
is a right triangle, or is an acute triangle according as it has an
obtuse angle, has a right angle, or all of its angles are acute.   In
a right triangle, the side opposite to the right angle is the hypotenuse.
A triangle with two congruent sides is isosceles.   A triangle with three
sides of the same length is equilateral.

Theorem 4

In $\triangle ABC$, $S[CA] \cong S[CB]$ if and only if $\angle CAB \cong \angle CBA$.   If $S[CA] \cong S[CB]$
then the line bisecting $\angle ACB$, the line through C and the midpoint of
$S[AB]$, and the line that is the perpendicular bisector of $S[AB]$ are
all the same line.

Corollary 4

If $A \neq B$, the line that is the perpendicular bisector of $S[AB]$ is
also the set of all points that are equidistant from A and B.

Theorem 5

One side of a triangle is greater than a second if and only if
the angle opposite to the first is greater than the angle opposite to
the second.   (The greater side "subtends" the greater angle.)

Corollary 5

The hypotenuse of a right angle triangle is greater than each of
the other sides.

Theorem 6

The sum of the lengths of each two sides of a triangle is greater
than the length of the third side.

Corollary 6

If $d(A,B) + d(B,C) = d(A,C)$, then the points A,B,C are collinear.
If they are also distinct, then $<ABC>$.

The following are the standard congruence theorems of elementary geometry.

Theorem 7

(Angle-side-angle)   If a correspondence of two triangles, or of a triangle with itself, is such that two angles and the side between them are respectively congruent to the two corresponding angles and the side between them, then the correspondence is a congruence of the triangles.

Theorem 8

(Side-side-side)   If a correspondence of two triangles, or of a triangle with itself, is such that the three sides of one triangle are respectively congruent to the corresponding sides of the other, then the correspondence is a congruence of the triangles.

In Euclidean geometry, our next theorem is a corollary of Th.7. Here it is an independent theorem.

Theorem 9

(Hypotenuse-angle)   In a correspondence of right triangles, if the hypotenuse of one corresponds and is congruent to the hypotenuse of the other and if one pair of corresponding acute angles are congruent, then the correspondence is a congruence of the triangles.

Theorem 10

(Hypotenuse-side)   In a correspondence of right triangles, if the hypotenuse of one corresponds and is congruent to the hypotenuse of the other, and if one pair of corresponding sides are congruent, then the correspondence is a congruence of triangles.

The following inequalities associated with pairs of triangles are also useful.

Theorem 11

If $\triangle ABC$ and $\triangle DEF$ are such that $d(B,A) = d(E,D)$, $d(B,C) = d(E,F)$, then $\angle ABC^0 > \angle DEF^0$ if and only if $D((A,C) > d(D,F)$.

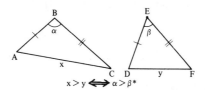

$$x > y \Longleftrightarrow \alpha > \beta *$$

We conclude this section with a somewhat special but useful definition for the distance between a point and a set.

Definition

(Foot in a set, distance between a point and set) Point P has
point F as a foot in set s if F belongs to s and if $d(P,F) \leqq d(P,X)$
for all X $\in$ s. If P has foot F in set s, then the distance between P
and s, denoted by $d(P,s)$, is the number $d(P,F)$.

Theorem 12

Each point of a set is its own unique foot in the set.

Theorem 13

Corresponding to a point P and a line t, there is exactly one line
through P that is perpenducular to t and the point at which this line
is perpendicular to t is the unique foot F in t of the point P.

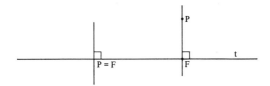

Corollary 13

If P has foot F in line t, and P $\neq$ F, then every point of L(PF)
has foot F in line t.

---

*
    We will sometimes use "=>" for "implies" and "<⟹" for "implies and
is implied by".

Theorem 14

If ∢PAB is acute, then the foot of P in L(AB) belongs to the open ray R(AB).

Section 4. Non-intersecting Lines

In Elliptic Plane Geometry, mentioned in I-5, lines are of limited extent and there are no non-intersecting lines. In Absolute, Euclidean, and Hyperbolic geometry, lines are of unlimited extent, and the weak exterior angle property (Th.3), which holds in all three geometries, impli the existence of non-intersecting lines. In fact, if P is not on line t, then - as we shall see - the family of lines through P must contain a non-intersector of t. But how many non-intersectors of t, one or

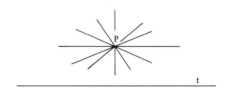

several? The axioms of Absolute geometry are not sufficient to decide the answer. If we add an axiom - that there is only one (the Playfair axiom, Th. D, I-3) - our geometry becomes Euclidean. If we assume - as we shall in the next chapter - that there is more than one, then our geometry becomes Hyperbolic.

In this section we consider some properties of non-intersecting lines in Absolute geometry, hence properties that are common to non-intersecting lines in both Euclidean and Hyperbolic geometry.

Definition

(Transversal of a set of lines)   A line t is a transversal of a set of lines if t is not in the set but intersects every line that is in the set.

Definition

    (Alternate interior and corresponding angles) If a line t is a
transversal of lines r and s at distinct points A and B then each of
the angles of r and t, and each of the angles of s and t, has a "transversal arm" contained in t and a "non-transversal arm" not contained in
t. An angle of r and t, and an angle of s and t, are alternate interior
angles if their transversal arms are opposite directed and intersecting
and their non-transversal arms lie in opposite closed sides of t. An
angle of r and t, and an angle of s and t are corresponding angles if
their transversal arms like directed and their non-transversal arms lie
in a closed side of t.

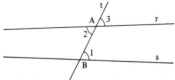

1 and 2 alternate interior—1 and 3 corresponding.

    Using Th.11 and Th.13 of II-2, it is a cumbersome but straight-
forward job to establish the following theorem.

Theorem 1

    If a line t is a transversal of lines r and s at distinct points
A and B, then a pair of alternate interior angles or a pair of corresponding angles are congruent if and only if each pair of alternate interior angles and each pair of corresponding angles are congruent.

    Our next result is just Th.B of I-3, whose proof was given there.

Theorem 2

    If line t is a transversal of lines r and s at distinct points A
and B respectively, and if a pair of alternate interior angles are congruent or a pair of corresponding angles are congruent then r and s are
non intersecting lines.

Corollary 2
     Two lines that are perpendicular to a third do not intersect.
Theorem 3
     If point p is not on line t then there is at least one line
through P that does not intersect **t**.
Proof
     Point P has a unique foot F in t and L(PF) ⊥ t, by Th.13, II-3.
There exists a line r that is perpendicular to L(PF) at P, Th.18, II-2.
Since r and t are both perpendicular to L(PF), they do not intersect
(Cor.2), hence r is one line through P that does not intersect t. □
     If a line separates two points then, by the plane separation
axiom (Axiom 4), it intersects their segment and hence their line. Thus
if two lines do not intersect, then neither separates any two points of
the other, which implies the following result.
Theorem 4
     If lines r and s are non-intersecting, then r is contained in one
side of s and s is contained in one side of r.
Corollary 4
     If r = L(AX), s = L(BY), and if X,Y are in one side of L(AB) then
R(BY) ⊂ In(⦠BAX).

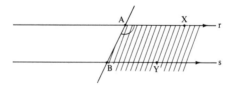

Proof
     Since R(BY) ⊂ s, then by Th.4 R(BY) ⊂ B-side of L(AX). By hypo-
thesis, Y ∈ X-side of L(AB) and so, by the edge-ray theorem (Th.2, II-2),
R(BY) ⊂ X-side of L(AB). Thus R(BY) ⊂ In (⦠ABX).   □
Definition
     (Strip between non-intersecting lines)   If lines r and s do not
intersect, the open strip between them is

          St(rs) = St(sr) = r-side of s ∩ s-side of r.

The union of lines r and s with the open strip between them is the <u>clo-</u>
<u>sed strip</u> between r and s, denoted by St[rs] or St[sr]. The lines r and
s are the <u>opposite</u> <u>edges</u> to both the open and closed strips.

<u>Definition</u>

(Set between non-intersecting lines)   A set S is <u>between</u> the
non-intersecting lines r and s if $S \subset$ St(rs).

<u>Theorem 5</u>

If r and s are non-intersecting lines and if A $\in$ r and B $\in$ s,
then a point of L(AB) is between A and B if and only if it is between
r and s.

<u>Proof</u>

By Th.4, A-side of s = r-side of s and B-side of r = s-side of r.
Since A $\in$ r and B $\notin$ r then, by the edge-ray theorem (Th.2, II-2),

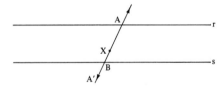

R(AB) $\subset$ s-side of r. Similarly, B $\in$ s and A $\notin$ s implies that R(BA) $\subset$
r-side of s. Therefore,

R(AB) $\cap$ R(BA) $\subset$ s-side of r $\cap$ r-side of s = St(rs).   (1)

Now if <AXB>, then, by Cor. 5.3, II-1, X $\in$ R(AB) $\cap$ R(BA). Thus, by
(1), <AXB> => X $\in$ St(rs).

Next, suppose that X $\in$ L(AB) $\cap$ St(rs). Let R(BA') be opposite
to R(BA). By the edge-ray theorem, s separates R(BA) and R(BA'), so
R(BA') $\subset$ non-r-side of s. Since X $\in$ r-side of s, then X $\notin$ R(BA'). Also,
X $\neq$ B, since B $\in$ s, so X $\notin$ R[BA'). This fact, with X $\in$ L(AB), implies
that X $\in$ R(BA). By a symmetric argument, X $\in$ R(AB), and so X $\in$ R(AB)
$\cap$ R(BA = S(AB), (Cor.5.3, II-1). Thus X $\in$ L(AB) $\cap$ St(rs) => <AXB>.   □

Theorem 6

If lines r and s are non-intersecting, then the non-s-side of r is a proper subset of the r-side of s.

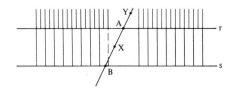

Proof

Consider Y ∈ non-s-side of r. Let B be a point of s. By the plane separation axiom (Ax.4), r ∩ S(YB) is a point A. By the edge-ray theorem R(BA) ⊂ r-side of s, and <BAY> implies that Y ∈ R(BA), so Y ∈ r-side of s. Thus the non-s-side of r is contained in the r-side of s. If X is a point of S(AB), then by Th.5, X ∈ r-side of s ∩ s-side of r. Thus the r-side of s contains points not in the non-s-side of r.   □

Corollary 6

If r = L(AV) and s = L(BW), with V,W in one side of L(AB), and if <BAY>, then In(≵YAV) is a proper subset of In(≵ABW).

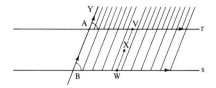

Proof

By hypothesis, the V-side of L(AY) is also the W-side of L(BA). By Th.6, the Y-side of L(AV) is a proper subset of the A-side of L(BW). Thus In(≵YAV) ⊂ In(≵ABW). But if X ∈ S(VW), then X ∈ In(≵ABW) and, by the proof of Th.6, X ∉ In(≵YAV). Thus In(≵YAV) is a proper subset of In(≵ABW).   □

Theorem 7

If no two of the lines r_1, r_2, r_3, intersect, and if line t is a transversal of these lines at A_1, A_2, A_3, respectively, then $<A_1A_2A_3>$ implies that r_2 is between r_1 and r_3 and that r_2 separates r_1 and r_3.

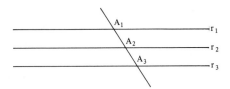

Proof

Since $<A_1A_2A_3>$, then, by Th.5, A_2 is between r_1 and r_3. Thus A_2 $\in$ r_3-side of r_1. By Th.4, all points of r_2 are in one side of r_1, hence r_2 $\subset$ r_3-side of r_1. By the same reasoning, A_2 $\in$ r_1-side of r_3 implies that r_2 $\subset$ r_1-side of r_3. Thus r_2 is contained in $St(r_1r_3)$. Since $<A_1A_2A_3>$, r_2 separates A_1 and A_3. Since A_1 and r_1 belong to the same side of r_2, and A_3 and r_3 belong to one side of r_2, the fact that A_1 and A_3 are in opposite sides of r_2 implies that r_1 and r_3 are in opposite sides of r_2. □

Section 5. Dedekind Cut, Continuity, A Basic Circle Property

In applying the ruler and protractor axioms to the absolute plane, great use is made of real number properties that are familiar from ordinary arithmetic. But we will need another and less familiar property that we will now describe rather informally.

Suppose that a and b are real numbers, with a < b, and that the closed interval $[a,b]$ = $\{x : a \leq x \leq b\}$ can be expressed as the union of two sets s_1, s_2 with the following properties.

(i) a $\in$ s_1 and b $\in$ s_2; (ii) $s_1 \cap s_2$ = $\emptyset$

(iii) x $\in$ s_1 and $<ayx>$ imply that y $\in$ s_1.

If x $\in$ s_1 then, by (iii), the whole interval $[a,x]$ is contained in s_1.

Now consider $z \in s_2$. If there existed w between z and b with w in s_1 then $[a,w] \subset s_1$ would imply that z was in s_1, and $z \in s_2 \cap s_1$ contradicts $s_1 \cap s_2 = \emptyset$. Thus if $z \in s_2$ then $[z,b] \subset s_2$. In this situation,

there are three possibilities. One is that s_1 is the singleton $\{a\}$ and s_2 is (a,b]. A second possibility is that s_1 is [a,b) and s_2 is the the singleton $\{b\}$. If neither of these occurs, then the successive classification of numbers as s_1-numbers or s_2-numbers must cause the right end of classified s_1-numbers and the left end of classified s_2 -numbers to 'move' toward each other. In all cases, it is a natural demand of continuity that there should exist some number c in [a,b] that marks the transition from set s_1 to set s_2 as x increases from a to b. Clearly, c must be either the biggest number in s_1 or the smallest number in s_2. The existence of the number c is the property of the real numbers we started out to describe.

In the situation described above, the sets s_1, s_2 are said to form "a Dedekind cut" of the interval [a,b] and the number c is the "Dedekind number" of the cut. Using metric coordinates for a line L(AB) one can translate the Dedekind cut of an interval [a,b] into the following conditions for the Dedekind cut of a segment S[AB].

Dedekind Cut Of A Segment

If the segment S[AB] can be expressed as the union of two sets s_1 s_2 with the following properties:

(i) $A \in s_1$ and $B \in s_2$;   (ii) $s_1 \cap s_2 = \emptyset$

(iii) $X \in s_1$ and <AYX> imply that $Y \in s_1$,
then there exists a unique point $C \in S[AB]$ such that <AXC> implies $X \in s_1$ and <CXB> implies $X \in s_2$.

The sets s_1, s_2 form the Dedekind cut of S[AB] and C is the Dede-

kind point of the cut. To illustrate the usefulness of such cuts, we
now apply the concept in deriving a circle property that we will need
later. First we introduce some definitions.

Definition

(Circle, interior, exterior)   Corresponding to a point A and a
positive number r, the circle with center A and radius r is the set
$$C(A,r) = \{X: \ d(A,X) = r\}.$$
The set of points interior to the circle is the set
$$In[C(A,r)] = \{X: \ d(A,X) < r\}.$$
and the set of points exterior to the circle is the set
$$Ex[C(A,r)] = \{X: \ d(A,X) > r\}.$$
If B ∈ C(A,r), then S[AB] is a radial segment and L(AB) is a radial
line.
In Euclidean geometry, the following property is usually obtained with
the help of the Pythagorean theorem, which we do not have (and which
does not hold in Hyperbolic geometry). We will, instead, employ a Dedekind cut.

Theorem 1

If a point P is interior to the circle C(A,r) then a line t through
P intersects the circle at exactly two points C,D and X on t is interior
to the circle if and only if <CXD>.

Proof

If t is a radial line then A ∈ t and there are exactly two points
on t at distance r from A (Th. 8, II-I), and A is between them. These
two points, C, D belong to t ∩ C(A,r), and clearly S(CD) = t ∩ In [C(A,r)].
Next, if t is not a radial line, let F be the foot in t of point
A, and let Q be one of the two points on t at distance r from F. By
the definition of a foot, d(A,F) ≤ d(A,P) < r. In the right triangle

Δ AFQ, the hypotenuse S[AQ] is greater than the side S[FQ], (Cor. 5, II-
so d (A,Q) > r.   Now define sets s_1 and s_2 as follows:

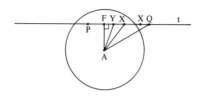

$$s_1 = \{X:\ X \in S[FQ] \text{ and } d(A,X) \leqq r\}$$
$$s_2 = \{X:\ X \in S[FQ] \text{ and } d(A,X) > r\}$$

Clearly, S[FQ]= s_1 U s_2 and s_1 ∩ s_2 = ∅.   Also, F ∈ s_1 and Q ∈ s_2.
Finally, consider X ∈ s_1 and <FYX>.   In the right triangles, Δ AFY and
ΔAFX, the angles ∡FYA and ∡FXA are acute.   But since <FYX>, ∡FYA and
∡XYA are supplements (Th. 11, II-2), hence ∡XYA is obtuse.   Thus ∡XYA^0
> ∡FXA^0 = ∡YXA^0.   Since a greater angle subtends a greater side (Th. 5,
II -3), d(A,X) > d(A,Y).   From d(A,Y) < d(A,X)< r, it follows that Y ∈ s
Thus the sets s_1, s_2 form a Dedekind cut of S[FQ] and a Dedekind point
C exists on S[FQ].   We will show that d(A,C) < r and d(A,C)> r are both
impossible, hence that C belongs to t ∩ C(A,r).

Case 1.

      d(A,C) < r.   Now, C ∈ s_1, so C ≠ Q.   The number h = ½ min{r-d(A,C),
d(C,Q)} is positive and there exists X on R(CQ) such that d(C,X) = h.
From d(C,X) < d(C,Q), it follows <CXQ>, hence X ∈ s_2.   Also, d(C,X) <

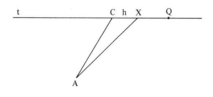

r − d(A,C).   Therefore, d(A,X) < d(A,C) + d(C,X) < d(A,C) + [r-d(A,C)]
= r implies that X ∈ s_1.Thus d(A,C) < r implies that s_1 ∩ s_2 ≠ ∅, so
d(A,C)< r is false.

Case 2

d(A,C) > r. Now C ∈ s_2 so C ≠ F. The number h = ½ min{d(A,C)-r,
d(C,F)} is positive and there exists X on R(CF) such that d(C,X) = h.
Since d(C,X) ≤ d(C,F), then <FXC>, which implies that X ∈ s_1. Also,

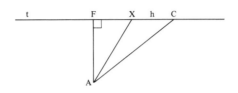

d(C,X) < d(A,C) - r. Therefore d(A,C) < d(A,X) + d(C,X) < d(A,X) +
[d A,C) - r] implies r < d(A,X) and hence that X ∈ s_2. Thus d(A,C) > r
implies that s_1∩ s_2 ≠ Ø , so d(A,C) > r is false.

From Case 1 and Case 2, it follows that d(A,C) = r, hence C ∈ t ∩
C(A,r). Point F is the midpoint of Q and a second point Q' on t at distance r from F. Repeating the argument for S[FQ'] that we used for S[FQ],
shows that there exists a point D on R(FQ') such that d(A,D) = r, hence
that {C,D} ⊂ t ∩ C(A,r).

Now let M, N be points of t such that M, D, C, N have that order

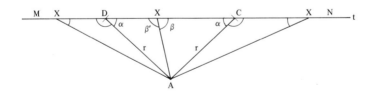

on t. Since d(A,C) = d(A,D) = r, then ∡ACD° = ∡ADC°= α and α < 90°(Cor.
3, II-3). If X ∈ S(CD), and if β = ∡AXC° and β' = ∡DXC°, then β + β'
= 180°. Thus either β ≥ 90°> α , and, from ΔAXC, d(A,C) = r > d(A,X),
or β' ≥ 90°> α and, from ΔAXD, d(A,D) = r > d(A,X). Thus S(CD) ⊂
In[C(A,r)]. If X ∈ R(DM), then <CDX> implies that ∡ADC and ∡ADX are
supplements. Since ∡ADC is acute, ∡ADX is obtuse, hence ∡ADX° > ∡AXD°,
which implies that d(A,X)> d(A,D) = r. Thus R(DM) ⊂ Ex[C(A,r)]. By an

entirely similar argument, R(DN) ⊂ Ex[C(A,r)]. Thus t intersects C(A,r)
only at C and D, and S(CD) = t ∩ In [C(A,r)].   ▫

Corollary 1.

        If r is a number greater than the distance of point A from line
t, then C(A,r) intersects t at two points C,D and the midpoint of S[CD]
is the foot of A in line t.

Section 6. Motions and Symmetries

        The natural symmetries of the absolute plane are expressed in
terms of distance preserving mappings called "motions of the plane". We
will not attempt a thorough study of this topic, but will simply develop
a few key ideas and properties that will be helpful later.

Definition

        (Isometry)   If Γ is a collection of ordered pairs of points
{(X,X')}such that (X,X') ∈ Γ and (X,Y') ∈ Γ implies X' = Y', then Γ
is a function or mapping. The set R of all first elements in the pairs
of Γ is the domain of Γ and the set S of all second elements is the range
of Γ , and R is said to be mapped onto S by Γ. If (X,X') is a pair in
Γ, then X' is the image of X, and this will also be denoted by X' = X Γ
and by X -> X'. If each point S is the image of exactly one point in R
then Γ is a "one-to-one", or "1-1" mapping.

It is a distance preserving mapping, or an isometry, if the distance between each two points of R is the same as that between their images in
S, i.e. d(X,Y) = d(X',Y') for all X,Y ∈ R.

Theorem 1.

        Every isometry is a 1-1 mapping.

Proof

        If X ≠ Y, then d(X,Y) > 0. Since d(X',Y') = d(X,Y), then

d(X',Y') > 0, hence X' ≠ Y'.   □

Definition

(Motion of a Set)   An isometric mapping of a set onto itself
is a _motion_ of the set.

If X -> X' is a motion Γ of a set _s_ the opposite correspondence
X' -> X is the inverse of Γ and is denoted by Γ⁻¹.   Since the same distances correspond under both Γ and Γ⁻¹, the following fact is immediate.

Theorem 2

The inverse of a motion is a motion.

Convention

If Γ is a motion of the plane and _s_ is any subset of the plane
then _s_Γ will denote the set consisting of the images of the points in
_s_.

Theorem 3

If X' = X Γ is a motion of the plane, then <ABC> if and only if
<A'B'C'>. The images of L(AB) and R(AB) and S[AB] are L(A'B') and R(A'B')
and S[A'B'] respectively. Non-collinearity is preserved. Each triangle
maps onto a congruent triangle and each angle maps onto a congruent angle.
If C is not in L(AB) then the C-side of L(AB) maps onto the C'- side of
L(A'B').

Proof

If <ABC> then A,B,C are three collinear points and d(A,B) + d(B,C)
= d(A,C). Since Γ is 1-1, A',B',C are three points, d(A,B) = d(A'B'),
d(B,C) = d(B',C'), and d(A,C) = d(A',C') imply that d(A'B') + d(B',C')
= d(A',C'), hence that <A'B'C'>.

Since Γ preserves betweeness, it is clear that R(AB)Γ is contained in R(A'B'). But if Y is a point of R(A'B'), it is the unique point
of the ray at distance d(A',Y) from A'. On R(AB) there is a unique point
X at distance d(A',Y) from A, thus X'must be Y. Therefore R(A'B') is
contained in R(AB). The opposite inclusions imply R(AB) = R(A'B').

Then L(AB) = R(AB) ∪ R(BA) must map onto R(A'B') ∪ R(B'A') = L(A'B').
Similarly, S[AB] = R[AB] ∩ R[BA] maps onto R[A'B') ∩ R[B'A') = S[A'B'].

If A,B,C are non-collinear points, the sum of each two of the
distances d(A,B), d(B,C), d(C,A) is greater than the third, so this must
also be true of the distances d(A',B'), d(B',C'), d(C',A'), hence
A', B', C' are non-collinear.

That △ ABC ≅ △ A'B'C' follows from the side-side-side congruence condition. This also implies that ∢ABC ≅ ∢A'B'C'.

If C is not in L(AB), then clearly C' is not in L(A'B'). If
X ≠ C and X ∈ C-side of L(AB), then X' ≠ C', and S[XC] ∩ L(AB) = ∅
implies that S[X'C] ∩ L(A'B') = ∅, hence that X' ∈ C'-side of L(A'B).
Thus the sides of L(AB) map into the sides of L(A'B'). But, if Y ≠ C'
and Y ∈ C'-side of L(A'B') there is, by hypothesis, some point Z that
maps onto Y. Since no point of L(AB)∪ non-C-side of L(AB) maps into
the C'-side of L(A'B'), then Z must be in the C-side of L(AB). Thus
the sides of L(AB) map onto the sides of L(A'B').   □

We now introduce some particularly useful motions of the absolute
plane.

Definition

(Identity mapping)   The mapping of the plane in which each
point is its own image is the identity mapping I. Thus XI = X for
all X.

Definition

(Reflection in a point)   The reflection of the plane in a point
A is defined as follows: point A is its own image, and if X ≠ A then
the image of X is the point X' such that A is the midpoint of S[XX'].
We denote the reflection in A by Γ_A.

Definition

(Reflection in a line)   The reflection of the plane in a line
t is defined as follows: if X ∈ t, then X is its own image, and if
X ∉ t then the image of X is the point X' such that t is the perpendicu-

lar bisector of S[XX']. We denote the reflection in t by Γ_t.

The identity mapping is obviously a motion of the plane, and we now show that reflections are also motions.

Theorem 4

Point reflections are motions of the plane.

Proof

Let X' = X Γ_A be the reflection of the plane in A. By definition A' = A. If Y $\neq$ A, then L(AY) exists and on L(AY) there are exactly two points Y,Z at distance d(A,Y) from A, and <YAZ>. Since A is the midpoint of S[YZ], Γ_A maps Y onto Z and Z onto Y. Thus Γ_A is a 1-1 mapping of A^2 onto itself.

Let B, C denote two arbitrary points in A^2. To show that Γ_A is an isometry, we must show that d(B,C) = d(B',C') for all B and C.

Case 1.

A $\in$ {B,C}, say A = B. By definition, d(A,C) = d(A' C'), and, since B' = B = A. this is also d(B,C) = d(B',C').

Case 2.

A $\notin$ {B,C} and A, B, C are collinear. Then <ABC>, <BCA>, or <CAB>. Consider <ABC>. Then C $\in$ R(AB), d(A,B) < d(A,C)

so $$d(B,C) = d(A,C) - d(A,B). \qquad (1)$$
From <BAB'> and <CAC'>, B' and C' are on the ray opposite to R(AB). Also, d(A,B') = d(A,B) < d(A,C) = d(A,C') implies that <AB'C'>
and so $$d(B'C') = d(A,C') - d(A,B). \qquad (2)$$
Because d(A,B) = d(A,B') and d(A,C) = d(A,C'), (1) and (2) imply that d(B,C) = d(B',C'). The argument for the other two subcases is entirely similar.

II-6

Case 3

A ∉ {B,C}and A,B,C are non-collinear. Because <BAB'> and <CAC'>, the angles ∢BAC and ∢B'AC' are congruent opposite angles. Since

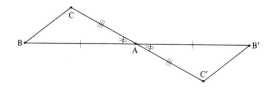

S[AB] ≅ S[AB'] and S[AC] ≅ S[AC'], then ΔBAC ≅ ΔB'AC'. by s-∢-s, so S[BC] ≅ S[B'C']as corresponding parts of congruent triangles.   □

Theorem 5

Line reflections are motions of the plane.

Proof

Let X' = X Γ_t denote the reflection of the plane in line t. If Y ∈ t then, by the definition of Γ_t,Y is its own image. If Y ∉ t, Y has foot F in t, and L(YF) is the unique line through Y and perpendicular t. On L(YF) there exactly two points Y,Z at distance d(F,F) from F and <YFZ>. Thus t is the perpendicular bisector of S[YZ] and so, by the definition of Γ_t, Y maps to Z and Z maps to Y. Thus Γ_t is a 1-1 mapping of A² onto itself.

Now let B,C denote two arbitrary points in A², and let F and G be their respective feet in t. To show that Γ_t is an isometry, we must show that d(B,C) = d(B',C') for all points B and C.

Case 1

Let L(BC)⊥ t. Then F = G and, by the definition of Γ_t, F is the midpoint of S[XX'] for X ∈ t and X ≠ F. Thus for all X ∈ t, $X\Gamma_t = X\Gamma_F$. Thus, from Th.4, we have

$$d(B,C) = d(B\Gamma_F,C\Gamma_F) = d(B\Gamma_t,C\Gamma_t) = d(B',C').$$

In all the remaining cases, the line L(BC)is not perpendicular to t

Case 2

The points B,C are both on t. Then B = B' and C = C' implies that d(B,C) = d(B',C').

Case 3

Just one of the points B,C is on t, B ∈ t. Then B' = B, foot
G is the midpoint of S[CC'] and t ⊥ L(CC').

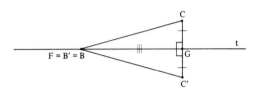

From S[GC] ≅ S[GC'], ∡BGC° = 90° = ∡BGC'° and S[GB] ≅ S[GB] , it follows
that ΔCGB ≅ ΔC'GB, by s-∡-s, and so S[BC] ≅ S[BC'] = S[B'C'] as corresponding parts of congruent triangles.

Case 4.

The points B,C lie in one side of t. Now L(FB) and L(GC) are
distinct lines perpendicular to t and hence are non-intersecting. By
the same argument as in Case 3, ΔCGF ≅ ΔC'GF, and so S[FC] ≅ S[FC'] and

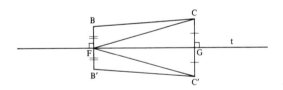

∡CFG ≅ ∡C'FG. By hypothesis, C ∈ B-side of t = L(FG). All points of
L(CG) are in one side of L(BF), so C ∈ G-side of L(BF). Thus, C ∈ In(∡BFG),
which implies that ∡BFC is the complement of ∡CFG. Since B' and C' are
both in the non-B-C-side of t, the same reasoning shows that C' ∈ In(∡B'FG),
and so ∡B'FC' is the complement of ∡C'FG. Since congruent angles have
congruent complements, and ∡CFG ≅ ∡C'FG, it follows that ∡BFC ≅ ∡B'FC'
Also, S[FB] ≅ S[FB'], hence Δ BFC ≅ ΔB'FC', by side-angle-side. Thus
S[BC] ≅ S[B'C'], as corresponding parts.

Case 5

The points B,C are in opposite sides of t. Now B and C' are in
one side of t and B' and C are in the other, and again L(FB) and L(GC)
are non-intersecting. Now the right triangles $\triangle$C'FG and $\triangle$CFG have
the congruence $\triangle$C'FG $\cong$ $\triangle$CFG, by side-angle-side, and so S[C'F] $\cong$ S[CF]

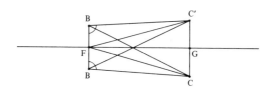

and $\angle$C'FG $\cong$ $\angle$CFG. Because B and C' are in one side of t, C' $\in$ In($\angle$BFG),
and $\angle$BFC' is the complement of $\angle$C'FG. Because B' and C are in one side
of t, C $\in$ In($\angle$B'FG), and so $\angle$B'FC is the complement of $\angle$CFG. Thus
$\angle$BFC'$\cong$ $\angle$B'FC (complements of congruent angles are congruent). Also,
S[FB] $\cong$ S[FB'], and so $\triangle$BFC' $\cong$ $\triangle$B'FC, by side-angle-side. As corresponding parts in this congruence, S[BC'] $\cong$ S[B'C] and $\angle$FBC'$\cong$ $\angle$ FB'C. Because <BFB'>, this last congruence is also $\angle$B'BC' $\cong$ $\angle$BB'C. Now, with
S[B'B] $\cong$S[BB'] , we have $\triangle$B'BC' $\cong$ $\triangle$BB'C, by side-angle-side, and S[B'C']
$\cong$ S[BC]as corresponding parts.   □

Definition

(Center of symmetry, line of symmetry) Point A is a <u>center of</u>
<u>symmetry</u>, or simply <u>a center</u>, to set s if s Γ_A = s. A line t is a <u>line</u>
<u>of symmetry</u>, or <u>axis of symmetry</u>, to set s if $s\Gamma_t$ = s.

From Th. 4 and Th.5 it follows that every point of A^2 is a center
of symmetry to the plane and every line of A^2 is a line of symmetry to the
plane.

The way in which a mapping relates various sets to their images
is called the "action" of the mapping. A point P is a <u>fixed</u> <u>point</u> of
a mapping Γ if PΓ = P. A set s is a <u>fixed</u> or <u>invariant</u> set of the mapping Γ if s Γ = s. The following facts about the action of point and line

reflections stem directly from the definition of these motions.

Theorem 6

In a point reflection Γ_A, the only fixed point is A. A line t
is a fixed line of the mapping if and only if it passes through A.
If A ∈ t, then Γ_A interchanges the opposite rays on t with origin A and
interchanges the sides of t. If A ∉ t, then t and t' = t Γ_A are non-
intersecting lines.

Theorem 7

In a line reflection Γ_t, a point X is a fixed point if and only
if X ∈ t. The fixed lines are t and the lines perpendicular to t. If
s ⊥ t at F, then Γ_t interchanges the opposite rays on s with origin F
and interchanges the sides of t but not the sides of s. If s does not
intersect t then its image s' does not intersect t or s; t lies in the
strip St(ss') and separates s and s'.

Because of the invariance of both distance and angle measure in
a motion, the following concept includes the congruence of triangles as
a special case.

Definition

(Congruent sets)   Sets R and S are congruent, denoted by
R ≅ S, if one is the image of the other in a motion of the plane.

In particular, then, every set is congruent to its image in a point
reflection or a line reflection. The identity I is a motion, and SI, = S,
for all S, so every set is congruent to itself.

Chapter III. Hyperbolic Plane Geometry

Introduction:

The properties in Chapter II belong to both absolute geometry and to Euclidean geometry, but the axioms there are sufficient to imply only a part of Euclidean geometry. For example, they do not imply that the angle sum of a triangle is 180°. To establish this, and many other facts of Euclidean geometry, some assumption equivalent to Euclid's "parallell postulate" is necessary. In traditional beginning geometry courses, the extra axiom most commonly adopted it the Playfair axiom:

"If P not on line t, then there exists exactly one line

through P that does not intersect t."

In developing a geometry different from Euclid's, Gauss, Bolyai, and Lobachevsky, all took as an axiom the assumption that through P there is more than one line that fails to intersect t. In this chapter, we shall do the same. A plane in which this particular denial of the Playfair acioms holds is called a "hyperbolic plane", and the basic properties we will deduce belong to hyperbolic geometry.

Since absolute geometry is an interesting structure in its own right, the notation "(A.G.)" following a theorem number will indicate that the theorem also holds in absolute geometry. The notation "Ex" following the statement of a theorem will indicate that the proof is left as an exercise.

Section 1. Hyperbolic Parallels

We now add the following axiom to the structure in Ch. II.

Axiom 9

(Hyperbolic Axiom)   If point P is not on line t, there are at least two lines through P that do not intersect t.

In talking about this new axiom, it will be convenient to have the following term.

<u>Definition</u>

(Pencil of lines)   The collection of all the lines that pass
through a point A is the <u>pencil of lines</u> at A, denoted by $P(A)$.

In the language of pencils, Ax. 9 states that if t does not belong to the pencil $P(P)$, then there are at least two lines of the pencil that do not intersect t.   There are several natural questions about
this novel situation.   Which lines of the pencil are the non-intersectors?
Are there more than two?   How are the intersectors and non-intersectors
of t distributed in the pencil?

The symmetries which the hyperbolic plane H^2 inherits from the
absolute plane A^2 give at once a partial answer to the questions just
raised.   If point P is not on line t, then P has a foot F in t.   Let
$u = L(PF)$ and let $X' = X\Gamma_u$ be the reflection of H^2 in u.   Because $t \perp u$,
t reflects onto itself, i.e. $t' = t\Gamma_u = t$.   The whole pencil $P(P)$ reflects
onto itself, but only two lines of the pencil are their own images,
namely u and the line r that is perpendicular to u at P.   If line s
in the pencil is not u or r, then $s' = s\Gamma_u \neq s$, and s and s' are interchanged by Γ_u.   Moreover, the motions of H^2 preserve intersection and
non-intersection.   Thus, $s \cap t \neq \emptyset \Rightarrow s' \cap t' \neq \emptyset \Rightarrow s' \cap t \neq \emptyset$ and
$s \cap t = \emptyset \Rightarrow s' \cap t = \emptyset$.   It follows that the intersectors of t occur

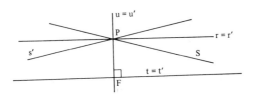

in pairs (except for u) that are symmetric with respect to u and the
same (except for r) is true of the non-intersectors.   In particular,
there is a non-intersector different from **r** (Ax. 9) which reflects to
a non-intersector, so $P(P)$ must contain at least three non-intersectors

of t. We can state this symmetric distribution in the following
way.

Theorem 1

(A.G.)   If point P is not on line t and if u is the line of
the pencil P(P) that is perpendicular to t, then u is an axis of symmetry to the set of lines in the pencil that intersect t and also to
the set of lines in the pencil that do not intersect t.

We want to obtain more precise information than in Th.1 and in
doing so the following language will be useful.

Definition

(Line subdividing an angle)   A line _subdivides_ an angle if it
passes through the angle vertex and intersects the angle interior.

The next theorem simply restates certain properties from absolute
geometry.

Theorem 2

(A.G.)   If X is interior to ⊁ABC, then L(BX) subdivides the
angle, L(BX) ∩ In(⊁ABC) = R(BX), and L(BX) separates R(BA) and R(BC).
(Ex.)

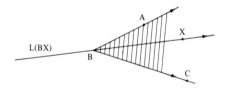

We can now establish the odd phenomenon that in this geometry
every line is contained in some angle interior.

Theorem 3

If point P is not on line t, then there exists an angle at P
whose interior contains t and whose bisector is perpendicular to t.

Proof

In the pencil P(P), let u be the line perpendicular to t and
let r be the line perpendicular to u. Line r is a non-intersector of

t and, by Ax.9, there is at least one other line in $P(P)$, say line s,
that does not intersect t. Because s is not perpendicular to u, two of
the opposite angles formed by s and u are acute. Thus, in s there
is a ray R(PA) such that $\sphericalangle FPA^{\circ} = \alpha < 90^{\circ}$, where $\{F\} = u \cap t$. By Th.1,
the line $s' = s\Gamma_u$ is also a non-intersector of t. If $A' = A\Gamma_u$, u
separates A and A' and $\sphericalangle FPA'^{\circ} = \alpha$. Because $\sphericalangle FPA$ is acute, the point G
that is the foot in u of A and A' belongs to R(PF), (Th.14,II-3), and
G is the midpoint of S[AA']. Clearly, u bisects $\sphericalangle APA'$, (Th.4, II-3)

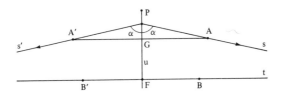

We want to show that $t \subset \text{In}(\sphericalangle APA')$. Toward this end, let B denote
a point on t in the A-side of u and let $B' = B\Gamma_u$. Because $G \in S(AA')$,
G is interior to $\sphericalangle APA'$, so $R(PG) \subset \text{In}(APA')$. But $R(PG) = R(PF)$,
because $G \in R(PF)$, so $R(PF) \subset \text{In}(\sphericalangle APA')$. By Th.7, II-2, this last
relation implies that

$$\text{In}(\sphericalangle FPA) \subset \text{In}(\sphericalangle APA'), \qquad (1)$$

and $\text{In}(\sphericalangle FPA') \subset \text{In}(\sphericalangle APA'). \qquad (2)$

Because $s \cap t = \emptyset$, and A,B are in one side of u, it follows from Cor.4,
II-4, that $R(FB) \subset \text{In}(\sphericalangle FPA)$. This, with (1), implies that $R(FB) \subset$
$\text{In}(\sphericalangle APA')$. Similarly, $s' \cap t = \emptyset$ and A',B' in one side of u imply that
$R(FB') \subset \text{In}(\sphericalangle FPA')$. This, with (2), shows that $R(FB') \subset \text{In}(\sphericalangle APA')$.
Thus we have,

$$t = R(FB) \cup \{F\} \cup R(FB') \subset \text{In}(\sphericalangle APA'). \qquad \square$$

Corollary 3.

Every line in the pencil at P that intersects t subdivides the
angle $\sphericalangle APA'$.   (Ex.)

As Cor.3 indicates, Th.3 gives new information about the lines
of $P(P)$ that are intersectors of t. However, there could be many angles
at P with the properties in Th.3. Our next theorem characterizes one
of these uniquely.

Theorem 4

If P is not on t, there is exactly one angle at P with the properties of Th.3 and such that every line subdividing this angle intersects t.

Proof

As in the proof of Th.3, let u be the line of $P(P)$ that is perpendicular to t, and let $\angle APA'$ denote an angle with the properties of
Th.3. Again, let B be a point on t and in the A-side of u and let
$B' = B\Gamma_u$. Also, let r denote the line of $P(P)$ that is perpendicular
to u, let C be a point on r in the A-B-side of u, with $C' = C\Gamma_u$.

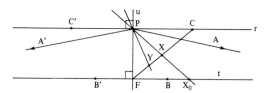

We can section the segment $S[FC]$ by sets s_1, s_2 defined as follows:

$$s_1 = \{X: \ X \in S[FC] \text{ and } L(PX) \cap t \neq \emptyset\}$$

$$s_2 = \{X: \ X \in S[FC] \text{ and } L(PX) \cap t = \emptyset\}.$$

If X is a point of $S[FC]$ then X is not P so $L(PX)$ exists and either does
or does not intersect t, so $S[FC] = s_1 \cup s_2$ To establish that s_1, s_2
form a Dedekind cut of $S[FC]$, we need to show (cf. II-5):

(i) $F \in s_1$ and $C \in s_1$ (ii) $s_1 \cap s_2 = \emptyset$;

(iii) $<FYX>$ and $X \in s_1$ imply that $Y \in s_1$.

Since $L(PF)$ intersects t at F, $F \in s_1$. Because $L(PC) = r$, and $r \cap t = \emptyset$,
$C \in s_2$. Thus (i) holds. No line is both an intersector and non-intersector of t, so (ii) holds. Finally, consider $X \in s_1$. and $<FYX>$. Because $X \in s_1, L(PX)$ intersects t at some point X_0. Since Y is between

F and X, L(PX) ≠ L(PF), and so X_0 ≠ F. Also, because X ∈ s_1, X ≠ C,

hence <FXC>. Because F and C are on opposite edges of the strip St[rt],

<FXC> implies that X is between r and t, (Th.5, II-4), which, in turn

implies <PXX$_0$>, since P and X_0 are on oppsite edges of St[rt]. Thus,

R(PX) = R(PX$_0$) and ∡FPX = ∡FPX$_0$. Since Y ∈ S(FX), Y ∈ In(∡FPX), and

so R(PY) ⊂ In(∡FPX) = In(∡FPX$_0$). By the Pasch properties of ΔFPX$_0$,

(Th.2,II-3), L(PY) intersects S(FX$_0$) and so intersects t. Thus Y ∈ s_1

and property (iii) holds. Therefore s_1 and s_2 do form a Dedekind cut

of S[FC] and the Dedekind point D of the section exists. Since both

L(PC) and L(PA) intersect S[FC] at points of s_2 D is not C. Also, the

line through P and any point of R(FB) intersects S[FC] at a point of

s_1 that is not F, so D ≠ F. Thus D is between F and C.

　　　For an indirect proof that the Dedekind point D belongs to s_2,

　　　　　　　　assume that D ∈ s_1,　　　　　　　　　(*)

hence that L(PD) intersects t at some point D$_0$. Because D is interior

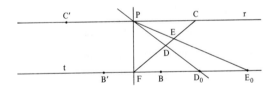

to ∡FPC, L(PD) subdivides ∡FPC, and, by Th.2, the intersection L(PD)

with In(∡FPC) is R(PD). On t, the ray R[FB') lies in the closed C'-side

of u = L(PF), so R[FB') ∩ In(∡FPC) = ∅. Therefore{D$_0$} = L(PD) ∩ t =

R(PD) ∩ R(FB). On R(FB), let E$_0$ be a point such that D(F,E$_0$) > d(F,D$_0$),

hence such that <FD$_0$E$_0$>. Because <FD$_0$E$_0$> and <FDC>, E$_0$ and C are in

the non-F-side of L(PD), so E$_0$ ∈ C-side of L(PD). All points of t

lie in one side of r, because r ∩ t = ∅, so E$_0$ ∈ D$_0$-side of L(PC).

Thus E$_0$ is interior to ∡D$_0$PC, so E$_0$ ∈ In(∡DPC) = In(∡D$_0$PC). Therefore

R(PE$_0$) intersects S(DC) at some point E. Because D is the Dedekind point,

$E \in S(DC)$ implies that $E \in s_2$ and hence that $L(PE) \cap t = \emptyset$. But $L(PE)$ intersects t at E_0, so $L(PE) \cap t \neq \emptyset$. The contradiction shows that the (*)-assumption cannot hold. Therefore $D \in s_2$ and $L(PD)$ is a non-intersector of t.

Now consider the reflection in u. The segment $S[FC]$ maps onto $S[FC']$, and D maps to D'. By Th. 1 and the properties of Γ_u it follows that $s_1\Gamma_u = s_1' = S[FD')$ and $s_2\Gamma_u = s_2' = S[D'C']$ form a Dedekind section of $S[FC']$ with Dedekind point D'. That is, $X' \in S[FD] \Rightarrow L(PX') \cap t \neq \emptyset$, and $X' \in S[D'C'] \Rightarrow L(PX') \cap t = \emptyset$. Because $\angle FPD$ is acute

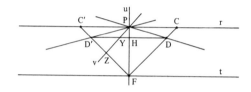

the foot in u of point D (and of D') is a point H that belongs to $R(PF)$. Thus $\angle HPD = \angle FPD$ and $\angle HPD' = \angle FPD'$. If a line v subdivides $\angle DPD'$. it intersects the interior of the angle and hence intersects $S(DD')$ at some point Y. If $Y \in S(D'H)$ then v subdivides $\angle HPD = \angle FPD$, so v intersects $S(FD')$ at a point $Z \in s_1'$ hence $v = L(PZ)$ intersects t. Similarly, if $Y \in S(HD)$ then v intersects $S(FD)$ at $Z \in s_1$, and $v = L(PZ)$ intersects t. Finally, if $Y = H$ then $v = L(PH)$ intersects t at F.

In all cases, then, if v subdivides $\angle DPD'$ it intersects t. The converse, that if v intersects t it subdivides $\angle DPD'$ follows from Cor. 3.

In conclusion, consider the triangle $\triangle FCC'$. Any line v in the pencil $P(P)$ intersects side $S[CC']$ of the triangle. By the Pasch property of the triangle, v must intersect a second side, hence must intersect $S[FC] \cup S[FC']$. If it intersects $S[FD)$ or $S[FD')$ it intersects t. If it intersects $S[DC]$ or $S[D'C']$, it does not intersect t. Thus every line in $P(P)$ is classified. □

Definition

(Fan angle of a point and line)   If point P is not on line t, the angle at P with the properties in Th.4 is the __fan angle__ of P and

t, and will be denoted by $\not\!\prec(P,t)$

To see what motivates the next step in our study, consider a

ray R(AB) and a point P not on the line L(AB). Instead of looking at

the lines of the pencil P(P), let us consider the family of all

open rays R(PX) at P, and let us search for those rays in the family

that have the following two properties:

(i) L(PX) ∩ L(AB) = ∅;

(ii) every ray between R(PX) and R(PA) intersects R(AB).

If our geometry were euclidean, only two rays would have property (i),

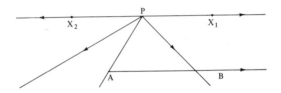

namely the opposite rays R(PX$_1$) and R(PX$_2$) on the line through P and

parallel to L(AB). Of these two rays, only the one in the B-side of

L(PA) has property (ii). Thus in euclidean geometry the conditions

(i) and (ii) determine a unique ray at P.

Now consider the same situation in our present geometry. By

Th.4, there exists a fan angle of point P and the line t = L(AB), say

$\not\!\prec$CPD, and one open arm of the angle, say R(PC) lies in the B-side of

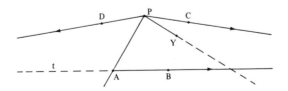

L(PA). We know that L(PC) ∩ L(AB) = ∅. Also, if R(PY) is between
R(PC) and R(PA) then L(PY) subdivides ≮CPD. Thus L(PY) intersects t
which clearly implies that R(PY) intersects R(AB). Thus R(PC) has
both the properties (i) and (ii). Next, consider a ray R(PX) that is
not R(PC). If R(PX) is on L(PA), the L(PX) intersects L(AB) and R(PX)
does not have the property (i). If R(PX) is in the D-side of L(PA)
and R(PY) is between R(PX) and R(PA) then L(PA) separates both R(PX)

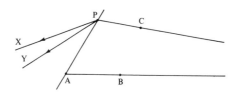

and R(PY) from R(AB). Thus R(PY) ∩ R(AB) = ∅, so R(PX) does not have
property (ii). If R(PX) is in the C-side of L(PA) then, by Th.9, II-2,
one of the rays R(PX) and R(PC) is between the other and R(PA). If
R(PX) is between R(PC) and R(PA) then, by property (ii) of R(PC),
R(PX) ∩ R(AB) ≠ ∅, so R(PX) does not have property (i). If R(PC)

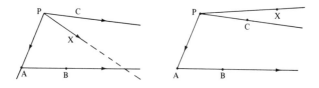

is between R(PX) and R(PA) then by property (i) of R(PC), R(PC) ∩ R(AB)
= ∅, so R(PX) does not have property (ii). Thus R(PC) is the only
open ray at P with both properties, and this fact is the motivation
for the following definition.

Definition

(Ray parallel to a ray)   An open ray R(PC) is <u>parallel to</u> an
open ray R(AB) if the lines of the rays do not intersect and if
R(PX) ⊂ In(∡APC) implies that R(PX) and R(AB) do intersect. A ray
R is parallel to a ray S, denoted by R ‖ S, if the open ray of R is
parallel to the open ray of S. That R is not parallel to S will be
denoted by R ‖̸ S.

From the argument preceding the definition, we have the following
uniqueness property.

Theorem 5

If point P is not on the line L(AB), there is exactly one closed
ray at P which is parallel to R(AB) and it is an arm of the fan angle
of P and line L(AB). If this ray is R[PC) then R(PC) is the only open
ray at P parallel to R(AB).

It is important to observe that the definition of one ray being
parallel to a second is not symmetric. Therefore we do not know as
yet that R(PC) ‖ R(AB) implies that R(AB) ‖ R(PC). To establish
this symmetry, along with several other basic properties of parallel
rays, is our next task. The theory is foundational in character and
some of the proofs involve a tedious carefulness about order relations.
But this is unavoidable if we are to truly justify this background
for the more lively mathematics of later sections.

Since the definition of parallel rays makes sense in absolute
geometry, the following is a theorem in that geometry.

Theorem 6

(A.G.)   If R(PQ) ‖ R(AB) then R(AB) ⊂ In(∡APQ).   (Ex.)

Theorem 7

(A.G.)   If point P is not on line t, and if B is between A
and C on t, then R(PQ) is parallel to R(AB) if and only if it is

parallel to R(BC).

<u>Proof</u>

Suppose first that R(PQ) || R(AB). Because L(PQ) ∩ L(AB) = ∅, and L(AB) = L(BC), then L(PQ) ∩ L(BC) = ∅. Next, consider

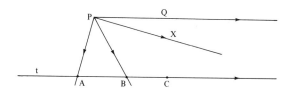

R(PX) ⊂ In(⊀BPQ). By Th.6, R(AB) ⊂ In(⊀APQ) and this implies that R(PB) ⊂ In(⊀APQ). Thus, from Th.7. II-2, we have R(PX) ⊂ In(⊀BPQ) ⊂ In(⊀APQ) so R(PX) intersects R(AB). Because L(PB) subdivides ⊀APQ it separates A and Q, and   because <ABC> it also separates A and C. Thus Q and C are in one side of L(PB) and R(BA) is in the other. Therefore R(PX), which is in the Q-side of L(PB), does not intersect R[BA]. Since it intersects R(AB), which is also S(AB] ∪ R(BC), R(PX) must intersect R(BC). Therefore R(PQ) || R(BC).

Next suppose that R(PQ) || R(BC). As before this implies that L(PQ) ∩ L(AB) = ∅. By Th.6. Q and C are in one side of L(PB). Since <ABC>, C and A are in opposite sides of L(PB). Therefore Q and A are in opposite sides of L(PB), and S(AQ) intersects L(PB) at a point D. Because A and Q are on opposite edges of the strip between the non-intersecting lines L(PQ) and L(AB), and <ADQ>, D is between L(PQ) and L(AB), (Th. 5, II-4). In turn, this implies that D on L(PB) is between P and B, (Th. 5, II-4). Thus R(PD) = R(PB), and

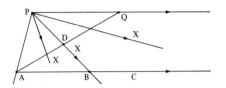

R(PD) ⊂ In(⊀APQ) implies R(PB) ⊂ In(⊀APQ). Therefore corresponding
to R(PX) ⊂ In(⊀APQ) there are three possibilities:

$$\text{(i) } R(PX) \subset In(⊀APB);$$

$$\text{(ii) } R(PX) = R(PB);$$

$$\text{(iii) } R(PX) \subset In(⊀BPQ).$$

In case (i), the Pasch properties of ΔAPB imply that R(PX) intersects
S(AB) and so intersects R(AB). In case (ii), R(PX) intersects R(AB)
at B.   In case (iii), R(PQ) || R(BC) implies that R(PX) intersects
R(BC). But R(BC) ⊂ R(AB), because <ABC>, so R(PX) intersects R(AB).
In all cases, then, R(PX) ∩ R(AB) ≠ ∅ hence R(PQ) || R(AB).   □

Corollary 7

If ray R(PQ) is parallel to a ray s in line t, then R(PQ) is
parallel to every ray in line t that has the same direction as s.   (Ex.)

Theorem 8

If point B is between points A and C on line s, then R(AB) is
parallel to R(DE) if and only if R(BC) is also parallel to R(DE).

Proof

Let t = L(DE). Suppose first that R(AB) || R(DE). Then
s = L(AB) does not intersect L(DE), so s = L(BC) does not intersect
L(DE). Now consider R(BX) ⊂ In(⊀DBC). By Th.7, II-2, R(BX) intersects

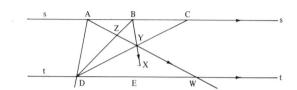

S(DC) at some point Y. Because <ABC>,L(BD) separates A and C. Since
R(BX) ⊂ C-side of L(BD), L(BD) separates A and R(BX) and therefore
separates A and Y. Thus S(AY) intersects L(BD) at a point Z. Now
we want to show that Y ∈ In(⊀DAB).

Clearly, R(AB) ⊂ B-side of L(AD). Because <ABC>, C is on R(AB)
and is therefore in the B-side of L(AD). By the edge-ray theorem,
(Th.2, II-2), R(DC) ⊂ B-side of L(AD) and therefore Y on R(DC) is in the
B-side of L(AD). Because Y is on R(BX), Y is in the D-side of L(BC) =
L(AB). Thus Y ∈ In(⊀DAB), and therefore R(AY) ⊂ In(⊀DAB). Because
R(AB) || R(DE), it follows that R(AY) must intersect R(DE) at some point
W.

The points C and D lie on the non-intersecting lines s and t
respectively,. Since <CYD>, Y is between s and t, (Th. 5 II-4), and
therefore Y on L(AW) is between A and W, (Th.5 II-4). From <AZY>
and <AYW> it follows that <ZYW>. Therefore the line of L(BX) intersects side S[ZW] of ΔDZW at Y. By the Pasch property of the tri-   ,
angle, (Th. 2, II-3) L(BX) must intersect S[ZD] ∪ S[DW]. But
L(BX) intersects L(DZ) at B. Because Z between A and W ⇒ Z between s and t, and this in turn implies <BZD>, B ∉ S[DZ]. Thus
L(BX) does not intersect S[ZD] and therefore must intersect S[DW].
The closed ray which is opposite to R(BX) lies in the non-t-side of
s and hence does not intersect t, so it does not intersect S[DW].

Therefore R(BX) intersects S[DW]. Since D is not on R(BX), R(BX)

intersects S(DW] and therefore intersects R(DE). Thus R(BC) ∦ R(DE).

Next suppose that R(BC) ∥ R(DE). As before, this implies that
L(AB) ∩ L(DE) = ∅. Let R[AX) be any ray interior to ≴DAB. We want to
show that R(AX) ∩ R(DE) ≠ ∅, hence that R(AB) ∥ R(DE).

Let X', B', D' be points such that <XAX'>, <BAB'>, <DAD'>.
The angle ≴D'AB' is opposite to ≴DAB, so R(AX') ⊂ In(≴D'AB'), (Th.8,
II-2). Therefore X' and B' are in one side of L(AD'). Because
<B'AB>, B'and B are in opposite sides of L(AD'). Therefore X' and B
are separated by L(AD'), and S(X'B) intersects L(AD') at some point Y.
Because X' and D' are in one side of L(AB') = s, and B ∈ s, the edge-
ray theorem (Th.2, II-2)implies that R(BX') ⊂ D'-side of s. Thus Y on

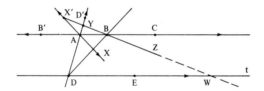

R(BX') is in the D'-side of s and so <DAY>.

Next, let R(BZ) denote the open ray opposite to R(BX'). Since
R(BX') ⊂D'-side of s, R(BZ) ⊂D-side of s = D-side of L(BC). The re-
lations, <B'AB> and <ABC>, imply that <B'BC>. Therefore L(BD) separates
B' and C and so separates R(BB') and R(BC). Because <B'AB>, A ∈ R(BB'),
so A ∈ non-C-side of L(BD). Therefore R(DA) ⊂ non-C-side of L(BD),
(Th,2, II-2). Since <DAY>, Y ∈ R(DA), so Y ∈ non-C-side of L(BD),
Therefore R(BY) ⊂ non-C-side of L(BD). Because <BYX'>, R(BY) = R(BX'),
so R(BX') ⊂ non-C-side of L(BD), which implies that R(BZ) ⊂ C-side of
L(BD). Similarly, since R(BX') ⊂ D'-side of s, R(BZ) ⊂ D-side of
s = D-side of L(BC). Thus R(BZ) ⊂ In(≴DBC), and this, with

R(BC) ∥ R(DE) implies that R(BZ) intersects R(DE) at some point W.

Now consider ΔDYW. The line L(AX) intersects side S[DY] at A and therefore intersects S[YW] ∪ S[DW], (Th.2, II-3). But L(AX) intersects L(YW) at X'. Since <X'YB> and <YBW> imply <X'YW>, X' is not on S[YW], and L(AX) ∩ S[YW] = ∅. Therefore L(AX) intersects S[DW]. The closed ray R[AX') lies in the non-t-side of s, hence does not intersect S[DW]. Therefore R(AX) intersects S[DW]. Because D is not on R(AX), R(AX) must intersect S(DW] and therefore must intersect R(DE) since S(DW] ⊂ R(DE). Therefore R(BC) ∥ R(DE). □

Corollary 8

If ray s is parallel to ray R(DE) then every ray in the line of s that has the direction of s is also parallel to R(DE).   (Ex.)

In conjunction with Th. 5, Cor. 8 implies a property that is not at all obvious, namely that if ⊁APB is the fan angle of P and line t then for each X in L(PB) one arm of the angle of X and t is contained in L(PB).

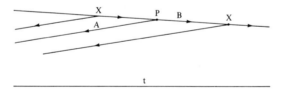

Our next proposition is rather obvious, but we state it for emphasis.

Theorem 9

If <ABC> and <DEF> and R(BC) ∥ R(EF), then R(BA) is not parallel to either R(EF) or R(ED), and R(ED) is not parallel to either R(BC) or R(BA).   (Ex)

We now make use of corollaries 7 and 8 to establish that parallelism of rays is a symmetric relation.

<u>Theorem 10</u>

        If R(PA) || R(QB) then R(QB) || R(PA).

<u>Proof.</u>

        Let F be the foot in L(QB) = t of point P, and let R(FC) be
the ray at F in t that is like directed R(QB). By Cor. 7, R(PA) ||
R(FC), so C ∈ A-side of L(PF) and L(FC) ∩ L(PA) = ∅.

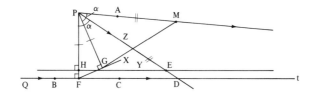

Next, let R(FX) be an arbitrary ray interior to ⅊PFC. We wish to
show that R(FX) must intersect R(PA), hence that R(FC) || R(PA).

        We first observe that ⅊PFX is acute, so G, the foot of P in
L(PX),must be in R(FX). Thus G is in the C-side of L(PF). If G is
on L(PA) or in the non-t-side of L(PA)then R(FG) = R(FX) obviously
intersects R(PA), so we suppose that G is in the t-side of L(PA).
Then G ∈ A-side of L(PF) ∩ F side of L(PA) = In(⅊FPA). Thus α = ⅊GPA°
< ⅊FPA°   so there exists R(PZ) between R(PF) and R(PA) and such
that ⅊FPZ° = α = ⅊GPA°. Because R(PA) || R(FC), R(PZ) intersects
R(FC) at some point D. Since G is the foot in L(FX) of point P, and
G ≠ F, d(P,G)< d(P,F). Thus if H is the point in R(PF) such that
d(P,H) = d(P,G), then <PHF>.

        Now let L(HY) denote the line perpendicular to L(PF) at H, with
Y in the A-C-side of L(PF). Because L(HY) and L(FC) are both perpendicular to L(PF) they do not intersect. In particular, L(HY)
does not intersect S[FD]. Since L(HY) intersects side S[PF] of ΔFPD,
it must intersect a second side, (Th. 2, II-3), and since it doesn't
intersect S[FD], it must intersect S[PD] at some point E which clearly
is neither P or D. On R(PA), let M be the point such d(P,M) = d(P,E).

This equality, with d(P,G) = D(P,H) and α = $\not\! <$GPM$^\circ$ = $\not\! <$HPE$^\circ$, implies

that ΔGPM $\cong$ ΔHPE. Thus, by corresponding parts of the congruence,

$\not\! <$PGM$^\circ$ = $\not\! <$PHE$^\circ$ = 90°. Thus L(GM) and L(FX) are both perpendicular

L(PG) at G, and hence are the same line. Therefore L(FX) intersects

R(PA) at M. Because M on L(FX) is in the A-side of L(FC), it must be

on R(FX). Therefore R(FX) intersects R(PA), which implies that

R(FC) || R(PA). Because R(FC) and R(QB) are like directed on t,

it follows from Cor. 8 that R(QB) || R(PA). □

Corollaries 7 and 8 and Theorem 9 and 10 allow us to extend
the concept of parallelism in the following rather natural way.

Definition

(Parallel ray and line, parallel lines)

A ray R is parallel to a line t if R is parallel to some ray

contained in t. A line s is parallel to a line t if some ray in s

is parallel to some ray in t. These parallelisms are symmetric

relations denoted by R || t or t || R and by s || t or t || s.

If s = L(AB) and t = L(CD) and if R(AB) || R(CD), then R(AB) is

said to be parallel to t in the direction R(CD) on t. Also, s

and t are said to be parallel in the direction R(AB) on s and

in the direction R(CD) on t.

Theorem 11

If point P is not on line t, there are exactly two lines through

P that are parallel to t. Each contains an arm of the fan angle

of P and t, and they are parallel to t in opposite directions

on t. (Ex.)

A basic property of parallelism that we have yet to consider

is transitivity. In euclidean geometry we know that if two lines

are parallel to a third then they are parallel to each other.

We can obtain a similar 'weak transitivity' in this geometry if we

make the added stipulation that the two lines are parallel to the

third in the same direction on the third. However, there is a difficulty that we must deal with first before we can give the proof that
we have in mind. In euclidean plane geometry, a transversal of three
parallel lines always exists. However, in hyperbolic geometry such
a transversal may or may not exist. To establish circumstances in
which it does is the purpose of the following lemma.

Lemma*

If two lines r and t are parallel to a third line s in the
same direction on s, then there is a line that intersects r, s
and t.

Proof

By hypothesis, r ∩ s = ∅, s ∩ t = ∅, and r ≠ t.
If r and t intersected at a point P, the fact that they are both
parallel to s in the same direction on s would contradict Th.11.
Therefore r ∩ t = ∅.

By hypothesis, there exist rays R(AX) ⊂ r, R(BY) ⊂ s, and
R(CZ) ⊂ t such that R(AX) || R(BY) and R(CZ) || R(BY). Let u = L(BC)
and consider the possible position of A. If A is on u then clearly
u intersects r, s and t. If A is in the non-t-side of s, L(AC) intersects r,s and t (because S(AC) intersects s). If A is in the non-s-
side of t, L(BA) intersects r, s, and t (because S(BA) intersects t).
There are two remaining possibilities:

(i)   A is between s and t and in the Y-Z-side of u;

(ii) A is between s and t and in the non-Y-Z-side of u.

In case (i), R(BA) ⊂ In(∢CBY) and R(BY) || R(CZ) implies that R(BA)
intersects R(CZ). Therefore L(BA) intersects r, s and t. In case
(ii), R(BC) ⊂ In(∢ABY) and R(BY) || R(AX) implies that R(BC) intersects

---

* A theorem that is established primarily to simplify the proofs
of other theorems is called a "lemma".

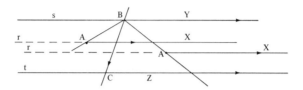

R(AX). Therefore L(BC) intersects r, s, and t. In all cases, then,
there is a transversal of r, s and t.   □

    We now have all the background we need to establish the follow-
'weak transitivity' theorem.

Theorem 12

    Two lines that are parallel to a third in the same direction
on the third are parallel to each other.

Proof

    Let r and t denote two lines that are parallel to a third
line s in the same direction on s. By the lemma, there exists a
line u that is a transversal of r, s and t at points A, B, C
respectively. Because r is parallel to s, it is parallel to one
of the two opposite open rays at B in s. Let R(BY) denote the ray
in s that is parallel to r. Since some ray in r is parallel to
R(BY), there is a like directed ray at A in r, say R(AX) that is parallel
to R(BY) (Cor. 8). Lines r and t are parallel to s in the same
direction on s, hence t ∥ R(BY) and there exists a ray R(CZ) in t that
is parallel to R(BY).
    Because R(AX) and R(CZ) are both parallel to R(BY), they both
lie in the Y-side of u. By hypothesis, r ∩ s = ∅. Thus A, B, C
are three points on u and exactly one of the relations <ABC>,
<BCA>, <CAB> holds. We will consider these possibilities separately
and show that in each case r is parallel to t.

Case 1   <ABC>.

Consider a ray R(AW) ⊂ In(✗CAX). Then L(AW) is not u and intersects

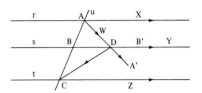

u only at A. Because <ABC>, A ∉ S[BC], and therefore L(AW) ∩ S[BC]
= ∅. Also, because <ABC>, R(AC) = R(AB) and ∗CAX = ∗BAX. Therefore
R(AW) ⊂ In(∗BAX) and this, with R(AX) || R(BY), implies that R(AW)
intersects R(BY) at some point D. Let A' and B' denote points such
that <ADA'> and <BDB'>.

Because <BDB'>, R(BD) and R(DB') are like directed. Since D ∈ R(BY),
R(BD) = R(BY). Therefore R(BY) and R(DB') are like directed, hence
R(BY) || R(CZ) implies that R(DB')|| R(CZ), (Cor.8).

Next, since <ADA'> and <ABC>, both A' and C are separated from
A by s, hence both lie in one side of s. Therefore R(DA') and R(DC)
are in one side of s. Because R(DA') ⊂ L(AW), and L(AW) ∩ S[BC] = ∅,
R(DA') is not R(DC) and is not between R(DB) and R(DC). Thus, by
Th.9, II-2, R(DA') is between R(DC) and R(DB'). Because R(DA') ⊂
In(∗CDB'), and R(DB') || R(CZ), R(DA') must intersect R(CZ). But
D ∈ R(AW) and <ADA'> imply that R(DA') ⊂ R(AW). Therefore R(AW)
intersects R(CZ), which implies that R(AX) || R(CZ) and hence that
r || t.

Case 2 <BCA>.

Since A is not on t, there exists a ray R(AV) that is parallel to
R(CZ). If r' = L(AV), we now have r' and s both parallel to R(CZ),
hence both parallel to t in the same direction on t. Since r' is
in the A-side of t and s is in the B-side of t, <ACB> implies that
t separates r' and s, hence r' ∩ s = ∅. By exactly the argument
of Case 1 it follows that R(AV) || R(CZ) and R(BY) || R(CZ) imply
R(AV) || R(BY). But, by Th.5, there is only one open ray at A

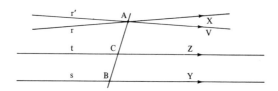

which is parallel to R(BY), and R(AX) || R(BY). Thus R(AV) || R(BY)
implies that R(AV) = R(AX), hence that r' = r. Therefore r' || t
is r || t.

Case 3   <CAB>

The argument in this case that r || t is entirely similar to that
Case 2 . (Ex.) □

Corollary 12

If two lines are parallel to a third in the same direction on
third, exactly one of the three parallel lines separates the other
two and it lies between the other two.   (Ex.)

Exercises - Section 1

1.   Prove Th. 2 (see theorems 4, 5 and 6 in II-2).

2.   Prove Cor. 3.

3.   Prove Th. 6 (See Cor. 4, II-4).

4.   Let R(AB) and R(CD) be two like directed rays in line t.
      Then either <ACD>, and R(AB) = R(AC), or else <CAB>, and
      R(CD) = R(CA). Use these properties, and Th. 7, to show
      that R(PQ) || R(AB) implies R(PQ) || R(CD). Thus prove
      Cor. 7.

5. Let R(AB) and R(CF) be two like directed rays in line s.
    Then either <ACF>, and R(AB) = R(AC), or else <CAB>, and
    R(CF) = R(CA). Use these properties and Th. 8 to prove
    that R(AB) parallel to R(DE) implies R(CF) || R(DE).
    Thus prove Cor. 8.

6. Prove Th. 9.

7. Let ✶APB be the fan angle of point P and line t. Let
    F be the foot in t of point P, and let R(FC) and R(FD)
    be opposite rays on t, with C in the A-side of L(PF).
    Explain why L(PA) || t in the direction R(FC) on t
    and why L(PB) || t in the direction F(FD) on t. If
    L(PX) is not L(PA) or L(PB), explain why L(PX) is not
    parallel to t. Thus explain Th. 11.

8. Prove Case 3 of Th. 12.

9. Explain how Cor. 12 follows from the Lemma and from Th.7,
    II-4.

10. Prove that if A,B,C are 3 collinear points, then R(AX) || R(BY)
    and R(BY) || R(CZ) imply R(AX) || R(CZ).

Section 2. Biangles, Hyperparallels

In this section we want to introduce one of the central figures in hyperbolic geometry and to establish some of its properties.

Definition

(Biangle)   The union of two closed, parallel rays with the closed segment joining their origins is a **biangle**. The closed rays are the **ray sides** of the biangle and the closed segment is the **segment side** of the biangle. Each endpoint of the segment side is the vertex of an angle that contains a ray side and the segment and these two angles are the **angles of** the biangle and their vertices are the **vertices of the biangle**. The notation (B-AC-D) will be used to denote the biangle whose angles are ⵁBAC and ⵁDCA and whose segment side is S[AC].   The interior of the biangle, denoted by

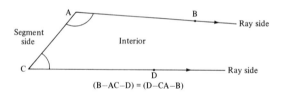

$(B-AC-D) = (D-CA-B)$

In(B-AC-D), is the intersection of the interiors of the two angles of the biangle.

Our first two theorems concern relations that we will refer to as "biangle Pasch properties" because they are analogs of the Pasch properties of triangles in absolute geometry.

Theorem 1

A line that intersects a biangle intersects at least two sides of the biangle or else is parallel to both ray sides and intersects the segment side between the vertices.

Proof

Let t denote a line that intersects the biangle (B-AC-D).
If a vertex is on t, say B ∈ t, then obviously t intersects the
segment side of S[BC] and the ray side R[BA), therefore we suppose
that neither A nor C is on t.   By hypothesis, there exists a point P
at which t intersects the biangle.   The point P is on a ray
side   or the segment side, and since P is neither A nor C, these
cases are exclusive and exhaust all possibilities.

Case 1.

P belongs to a ray side.   Consider P ∈ R[AB).   Since A ∉ t,
P ∈ R(AB) and t ≠ L(AB).   We may suppose that <APB>.

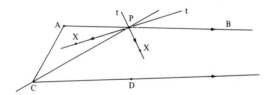

Let R(PX) denote the ray on t in the C-D side of L(AB).   Since
C ∉ t, R(PX) ≠ R(PC), so either R(PX) ⊂ In(∢APC) or R(PX) ⊂ In(∢BPC).
In the first instance, R(PX) intersects S(AC), by Th.7, II-2,
and in the second instance, R(PB) || R(CD) implies that R(PX) inter-
sects R(CD).   Thus t intersects two sides of the biangle.   The
argument for P ∈ R[CD) is entirely similar.

Case 2.

P belongs to the segment side.   Since neither A nor C is on
t, t ≠ L(AC), and P ∈ S(AC).   Now let R(PX) be the ray on t in the
B-D-side of L(AC) and let R(PY) denote the open ray at P parallel
to R(AB) and hence also to R(CD), (Th. 12, III-1).   If R(PX) ≠ R(PY)
then either R(PX) ⊂ In(∢APY) or R(PX) ⊂ In(∢CPY).   In the first

instance, R(PY) || R(AB) implies that R(PX) intersects R(AB), and
in the second, R(PY) || R(C) implies that R(PX) intersects R(CD).
Thus, t intersects two sides of the biangle. Finally, if R(PX) =
R(PY) then clearly t = L(PX) is parallel to both ray sides.   □

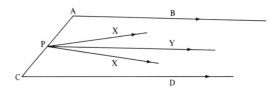

By an argument similar to that for Th. 1, one can establish
the following proposition.

Theorem 2

If a point P is interior to a biangle, then a line through
P intersects two sides of the biangle or else is parallel to both
the ray sides and intersects the segment side between the vertices.
(Ex.)

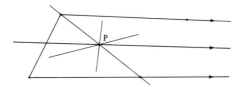

Like a triangle, a biangle has certain associated exterior
angles.

Definition

(Exterior angle to a biangle)   An angle that is adjacent
to an angle of a biangle and supplementary to it is an exterior
angle to the biangle at the vertex it shares with the biangle.

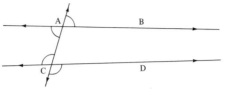

The four exterior angles to (B—AC—D).

We want to establish an exterior angle property of biangles that is similar to that of triangles. In doing so, we will need to make to make use of pairs of lines that are neither intersectors nor parallels. Such lines will play an important role throughout our study. At this point we merely wish to introduce a name for them and to derive certain of their properties which we can exploit in connection with biangles.

Definition

(Hyperparallels)   Two lines are <u>hyperparallels</u>, and each is hyperparallel to the other, if they do not intersect and are not parallel. That lines r and s are hyperparallel will be denoted by r)(s or s)(r and the negation of this will be denoted by r)⧸(s.

Our first proposition about hyperparallels follows immediately from the nature of fan angles and from a familiar property of both absolute and euclidean geometry, namely that if two lines intersect, the two lines that bisect their angles are perpendicular, (see exercises).

Theorem 3

Two lines that are perpendicular to a third line are hyperparallel to each other.

Proof

Let r and t denote two lines that are perpendicular to a line u at points P and F respectively. Since r ≠ t, P ≠ F. Let the fan angle of P and t be denoted by ⊀APB, and let the opposite angle be ⊀A'PB', with <APA'>. By the fan angle theorem, Th.4, III-1, the bisector

of these angles is the line through P that is perpendicular to t,

hence is the line u. Because line r in the pencil P(P) is per-

pendicular to u, it is the bisector of the opposite angles ∗APB'

and ∗A'PB. Because r does not subdivide the fan angle, it does not

intersect t, and since it bisects ∗APB', it is neither L(PA) nor

L(PB), and hence is not parallel to t. Therefore r)(t.   □

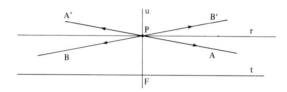

If point P is not on line t, we can make use of Th. 3 to

associate each intersector of t in the pencil P(P) with a particu-

lar hyperparallel in the pencil.

Theorem 4

If point P is not on line t, then corresponding to each

point Y in t there is a line s in the pencil P(P) such that s and

t are hyperparallel and L(PY) as a transversal of s and t forms

with them congruent alternate interior angles.

Proof

Let F be the foot in t of point P, let u = L(PF), and

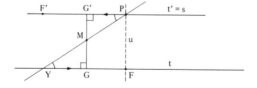

consider first a point Y ∈ t, Y ≠ F. Corresponding to M, the mid-

point of S[PY], let $X' = X\Gamma_M$ be the reflection of the plane in M.
The mapping leaves M fixed and interchanges Y and P. Thus t maps
onto a line t' in the pencil $P(P)$. The line L(PY) is fixed and
its sides are interchanged, (Th.6, II-6), hence L(PY) separates
R(YF) and R(Y'F') = R(PF'). Thus the angles ∡PYF and ∡YPF' are
alternate interior angles with respect to L(PY) as a transversal
of t and t'. They are congruent angles because Γ_M is a motion
of the plane and preserves angle measure, (Th.3, Th.4, II-6).

The angle ∡PYF is an acute angle in the right triangle ΔPYF.
Since M is on R(YP), its foot in t is a point F in R(YF), hence
$G' = G\Gamma_M$ belongs to R(PF'). Because $90° = ∡MGY° = ∡M'G'Y'° = ∡MG'P°$, and <GMG'>, if follows that t and t' are perpendicular to
L(GG'). Thus, by Th.3, t' and t are hyperparallel, so t' is a line
s with the properties of the theorem.

Finally, if Y = F and M is the midpoint of S[PF], then Γ_M
maps t to the line s in $P(P)$ that is perpendicular to u. Thus
s and t are hyperparallel, and the alternate interior angles that
u forms with s and t are congruent right angles.   □

Using the same sort of argument as in the last proof, one
can establish the following proposition.

Theorem 5

If a line v is a transversal of two lines s and t at distinct
points, and if it forms with s and t congruent alternate interior
angles then s and t are hyperparallel.   (Ex.)

We now return to biangles and will apply Th.4 to establish
an exterior angle property for such figures.

Theorem 6

An exterior angle to a biangle is greater than the non-adjacent angle of the biangle.

Proof

     Corresponding to a biangle (A-PC-D), let R(PA') denote the
open ray opposite to R(PA). The angle ⦨A'PC is exterior to the
biangle at P and we wish to show that ⦨A'PC° > ⦨PCD°.

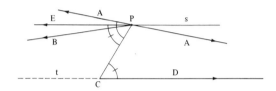

     Let t = L(CD) and let ⦨APB denote the fan angle of P and t.
By Th. 5, corresponding to C on t there exists a line s in the
pencil P(P) that is hyperparallel to t and such that L(PC) forms
congruent alternate interior angles with respect to s and t.
Thus there exists a ray R(PE) in the A'-B-side of L(PC), such that
⦨EPC° = ⦨DCP, and L(PE) = s. But because s is hyperparallel to t,
s is not L(PA) or L(PB) and s does not subdivide ⦨APB. Thus s
must subdivide ⦨A'PB, and R(PE) is between R(PB) and R(PA').
Thus ⦨A'PC° > ⦨EPC° = ⦨DCP°.   □

Corollary 6.1

     The sum of the measures of the two angles in any biangle is
less than 180°.   (Ex.)

Corollary 6.2

     If one angle of a biangle is a right angle, the other angle
of the biangle is acute.

     Because of Cor. 6.1, biangles are classified in the same
manner as triangles. A biangle is "obtuse" if it has an obtuse
angle, it is a "right" biangle if it has a right angle, and is an
"acute" biangle if both of its angles are acute.

     Since a biangle has only two angles and one measurable side,
we can define congruence for biangles more simply than for triangles.

Definition

(Congruent biangles)   Two **biangles** are **congruent** if their segment sides are congruent and the pair of angles in one biangle are congruent in some order to the pair of angles in the second biangle. The notation $(A-BC-D) \cong (E-FG-H)$ will be used to indicate that $\angle ABC \cong \angle BFG$, $S[BC] \cong S[FG]$, and $\angle DCB \cong \angle HGF$.

Theorem 7

(Segment-angle)   If two biangles have congruent segment sides, and if an angle in one is congruent to an angle in the other, then the remaining pair of angles are congruent.

Proof

Let $(A-BC-D)$ and $(E-FG-H)$ denote biangles such that $x = d(B,C) = d(F,G)$, $\alpha = \angle ABC^{\circ} = \angle EFG^{\circ}$, $\beta = \angle DCB^{\circ}$, and $\gamma = \angle HGF^{\circ}$.

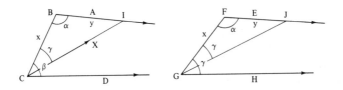

If $\beta \neq \gamma$, then one of the numbers β, γ is smaller. Suppose that $\gamma < \beta$. Then if $R(CX)$ in the A-D-side of $L(BC)$ is such that $\angle BCX^{\circ} = \gamma$, it follows from $\gamma < \beta$ that $R(CX) \subset In(\angle BAD)$, and $R(CD)$ $|| R(BA)$ implies that $R(CX)$ intersects $R(BA)$ at some point I. If $y = d(B,I)$ there is a point J on $R(FE)$ such that $d(F,J) = y$. Since both $\triangle CBA$ and $\triangle GFJ$ have sides of lengths x and y including an an-gle of measure α, $\triangle CBA \cong \triangle GFJ$, and hence $\angle FGJ^{\circ} = \angle BCI^{\circ} = \gamma$. But $J \in In(\angle FGH)$ implies that $\angle FGJ^{\circ} < \angle FGH^{\circ}$, so $\angle FGJ^{\circ} < \gamma$. Thus $\gamma < \beta$ is impossible. Similarly $\beta < \gamma$ is impossible. Thus $\beta = \gamma$. □

Theorem 8

(angle-angle)   If the two angles of one biangle have the

same measures as the two angles of a second biangle, then
the segment sides of the biangles are congruent.

Proof

Let (A-BC-D) and (E-FG-H) be biangles such that α = ⧸ABC° =
⧸EFG°, β = ⧸BCD° = ⧸FGH°, x = d(B,C), and y = d(F,G). On
R(CB) let P be the point such that D(C,P) = y, and let R(PX) be
the ray at P parallel to R(CD). By Th.7, (E-FG-H) ≅ (X-PC-D),

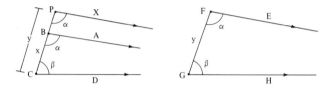

therefore ⧸CPX° = ⧸GFE°= α. If x ≠ y, then R(PX) ≠ R(BA), and since
R(PX) and R(BA) are non-collinear, and both are parallel to R(CD),
they are parallel to each other. Thus the biangle (X-PB-A) exists.
If x < y, then <CBP> implies that ⧸CBA is exterior to (X-PB-A)
and so, by Th.6, ⧸CBA° > ⧸BPX° = ⧸CPX°, or α > α, which is contradictory. If x > y, then <CPB> implies that ⧸CPX is exterior to
(A-BP-X), and ⧸CPX° > ⧸CBA° again asserts that α > α, which is
contradictory. Thus x and y must be equal, and S[BC] ≅ S[FG].   □

Exercises - Section 2

1. Prove Th. 2

2. Given ∢ABC and point D such that <ABD>, if R(BX) bisects
    ∡ABC and R(BY) bisects ∢CBD, show that L(BX) ⊥ L(BY).
    Thus prove that if two lines intersect, the bisectors of
    their angles are perpendicular (A.G.)

3. Prove Th. 5. Show that if s = L(AB), t = L(CD) and if
    L(BC) = v separates A and D, then s and t cannot inter-
    sect and cannot be parallel.

4. Prove Cor. 6.1.

5. A biangle is defined to be isosceles if its two angles are
    congruent. If (A-BC-D) is an isosceles biangle, and M is
    the midpoint of S[BC], there is a line s through M and
    parallel to R[BA) and R[CD), and there is a line t perpen-
    dicular to L(BC) at M. Prove that s = t. Thus prove the
    theorem: "The perpendicular bisector of the segment side
    of an isosceles biangle is parallel to both the ray sides."

6. Let (A-BC-D) be a biangle and let t be the perpendicular
    bisector of S[BC]. Prove that if t is parallel to the
    ray sides R[BA) and R[CD) then the biangle is isosceles.

7. If (A-BC-D) is an isosceles biangle whose congruent angles
    have measure α and if ΔPQR is an isosceles triangle whose
    congruent angles at P and Q have measure β, show that
    d(B,C) = d(P,Q) implies that β < α.

Section 3. Saccheri and Lambert Quadrilaterals, Polygon Angle Sums

In this section we shall see why the quadrilaterals associated with the names "Saccheri" and "Lambert" play a basic role in
hyperbolic geometry. Since we did not deal with quadrilaterals in
Chapter II, we begin with a definition of polygons that also holds
in the absolute plane.

<u>Definition</u>

(Circuits, polygons, convex polygons)   If n $\geq$ 3 and if
n points in the plane are assigned an order, then each segment that joins a point and its successor, and the segment joining
the last point to the first, are the segments or the <u>sides</u> of a
of a <u>circuit</u>, and the n points are the <u>vertices</u> of the circuit.
The circuit is a <u>polygon</u> if no two of the sides intersect except
at a vertex and each vertex is the intersection of exactly two
sides. The circuit is a <u>convex</u> <u>polygon</u> if it is a polygon and if
the line of each side does not separate any two vertices. A convex polygon is a <u>proper</u> convex polygon if no three of the vertices
are collinear. A proper, convex polygon with n vertices is also
a proper, convex <u>n-gon</u>.

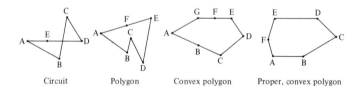

Circuit   Polygon   Convex polygon   Proper, convex polygon

Convention:

 Since we will work almost exclusively with proper, convex
polygons, it will be convenient to have the understanding that
the terms "polygon" and "n-gon" refer to a proper convex polygon
and n-gon respectively, unless something is said to the contrary.
In particular, the term "quadrilateral" will refer to a proper,
convex 4-gon. We will use the notation $P_1 P_2 ... P_n$ to represent
an n-gon, with vertices $P_1, P_2, ..., P_n$, corresponding to the
n-tuple order $(P_1, P_2, ..., P_n)$. Clearly, $(P_1, P_2, .., P_n)$ and $P_2 P_3 ... P_n P_1$
are representations of the same n-gon.

Definition

 (Parts of an n-gon).   Two sides of an n-gon $P_1 P_2 ... P_n$
are adjacent sides if their intersection is a vertex.
An angle is an angle of the n-gon if it contains two adjacent sides.
Two angles of the n-gon are neighbors if the segment joining their
vertices is a side. A segment that joins two vertices that are
not neighbors is a diagonal. An angle that is adjacent to an angle
of the n-gon, and supplementary to it, is an exterior angle to
the n-gon. Each side of the n-gon determines a line that is the ⸗
edge of an open half-plane containing the other n-2 vertices, and
the intersection of these n open half-planes is the interior of
the n-gon and will be denoted by $In(P_1 P_2 ... P_n)$. In the special
case of a quadrilateral, two sides that do not intersect are opposite sides, and two angles that are not neighbors are diagonally
opposite angles.

 We now introduce the quadrilaterals that played an important
role in the history of hyperbolic geometry.

Definition

 (Lambert quadrilateral)   A quadrilateral with three right
angles is a Lambert quadrilateral or L-quadrilateral.

Definition

(Saccheri-type quadrilateral)   A quadrilateral with two
neighboring right angles is a <u>Saccheri-type</u>, or <u>S-type</u> quadrilateral. With the segment joining the vertices of these two
right angles designated as the "base", the opposite side is
called the "summit" and the remaining two sides are called "the
sides" of the quadrilateral. Each of the angles containing the
summit and a side is a "summit angle", and a summit angle is said
to be "opposite" to the side it does not contain.

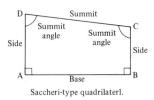

Saccheri-type quadrilaterl.

Definition

(Saccheri quadrilateral)   A <u>Saccheri</u> <u>quadrilateral</u> is
an S-type quadrilateral in which the two sides are congruent.
The segment joining the midpoints of the base and summit is the
<u>altitude</u> of the Saccheri quadrilateral, or S-quadrilateral.

If A $\neq$ B, there exist right angles $\angle$ABC and $\angle$BAD such that
C and D lie in the same side of L(AB). It is a straightforward
matter, though somewhat tedious, to verify that the circuit corresponding to the order ABCD is a proper, convex quadrilateral,
hence that S-type and S-quadrilaterals exist. Similarly, for
Lambert quadrilaterals, if C is interior to a right angle $\angle$XAY,
then the foot of C on L(AX) is a point B on R(AX), and the foot
of C on L(AY) is a point D on R(AY). Again, it can be verified
that the circuit corresponding to the order ABCD is a proper,
convex quadrilateral with three right angles hence is an L-quadrilateral.

The following theorem gives an obvious but useful property
of quadrilaterals and its proof serves as an example of how basic
properties of polygons can be established.

Theorem 1. (A.G.)

A point that belongs to a quadrilateral but not to an angle
of the quadrilateral is interior to that angle.

Proof

Let P be a point of the quadrilateral ABCD such that
P ∉ ∡DAB. No three of the vertices are collinear, so neither C
or D is on L(AB). Since L(AB) does not separate C and D, they
lie in one side of L(AB), hence C ∈ D-side of L(AB).

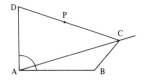

By the same reasoning, C ∈ B-side of L(AD), hence C ∈ In(∡DAB).
Thus P = C implies P ∈ In(∡DAB). If P ≠ C, then P ∈ S(BC) ∪ S(DC).
But C ∈ In(∡DAB) implies that In(∡BAC) ∪ In(∡DAC) ⊂ In(∡DAB),
hence that S(BC) ∪ S(DC) ⊂ In(∡DAB), and thus that P ∈ In(∡DAB).   □

Theorem 2   (A.G.)

The summit angles of a Saccheri quadrilateral are congruent
and the altitude line is perpendicular to both the summit line and
the base line.

Proof

Let ABCD be a Saccheri quadrilaterla with right angles at A
and B, with base S[AB], and with S[AD] ≅ S[BC]. Let M be the
midpoint of the base and N the midpoint of the summit S[CD].
Then ΔMAD ≅ ΔMBC, by side-angle-side, so α = ∡MDA° = ∡MCB°, and
S[MD] ≅ S[MC]. The summit S[CD] is thus the base of the isosceles

triangle ΔMDC, hence β = ✳MDC° = ✳MCD°. Since M ∈ S(AB), then,

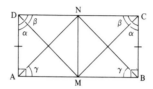

by Th. 1, M ∈ In(✳ADC) and M ∈ In(✳BCD). Therefore ✳ADC° = ✳ADM°
+ ✳MDC° = α + β . Similarly, ✳BCD° = ✳BCM° + ✳MCD°= α + β.
Thus the summit angles are congruent. Since the line L(MN) passes
through the vertex M and the midpoint of the base in the isosceles
triangle ΔMCD, L(MN) ⊥ L(CD), (Th.4, II-3). Because the summit
angles, ✳ADN and ✳BCN, are congruent, and S[ND] ≅ S[NC],
ΔADN ≅ ΔBCN, by side-angle-side, and therefore S[AN] ≅ S[BN].
This congruence implies that ΔANB is isosceles with base S[AB].
Since L(MN) passes through the vertex N of this triangle, and also
passes through the midpoint of the base, L(MN) ⊥ L(AB).   □

Corollary 2

        The base and summit lines of a Saccheri quadrilateral are
hyperparallel, and each side line is hyperparallel to the other
and to the altitude line.

Theorem 3. (A.G.)

        In an S-type quadrilateral, if the sides are not congruent
then the summit angles are not congruent and the greater summit
angle is opposite to the greater side.   (Ex.)

        The following converse to Th. 2 and Th. 3 also holds.

Theorem 4 (A.G.)

        If ABCD is an S-type quadrilateral, with summit angles at
C and D, then d(A,D) is greater than, equal to, or less than d(B,C)
according as ✳C° is greater than, equal to, or less than ✳D°.   (Ex.)

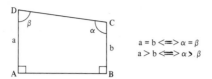

$$a = b \Longleftrightarrow \alpha = \beta$$
$$a > b \Longleftrightarrow \alpha > \beta$$

Our next theorem is Saccheri's famous acute angle hypothesis, discussed in Section 3 of Chapter I. Saccheri believed that he had proved that this property contradicts the axioms of absolute geometry. His proof is known to be incorrect, and in a later chapter we will discuss why his conclusion is also thought to be incorrect.

Theorem 5

The summit angles of a Saccheri quadrilateral are acute.

Proof

Let ABCD be a Saccheri quadrilateral with base S[AB], and with $\angle C^{\circ} = \angle D^{\circ} = \alpha$. Let points A' and D' be such that <ABA'> and <DCD'>, and let R(DX) be the open ray at D and parallel to R(AB), and let R(CY) be the open ray at C and parallel to R(BA').

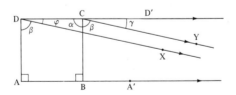

The right biangles (X-DA-B) and (Y-CB-A') are congruent, by segment-angle, (III-2, Th.8), hence $\angle ADX^{\circ} = \angle BCY^{\circ} = \beta$. Since the summit line L(DC) is hyperparallel to L(AB) at D, and R[DX] is an arm of the fan angle, $\angle(D, L(AB))$, R(DX) $\subset$ In($\angle ADC$). Similarly, R(CY) $\subset$ In($\angle BCD'$). Thus if $\varphi = \angle XDC^{\circ}$ and $\gamma = \angle YCD'^{\circ}$, then

α = ⦣ADC^{o} = β + φ and ⦣BCD'^{o} = β + γ. Because <ABA'>, R(AB) and
R(BA') are like directed, hence R(CY) || R(AB) || R(DX), and
the biangle (Y-CD-X) exists. Because ⦣YCD' is an exterior angle to
this biangle, γ > φ, (III-2, Th. 7), and therefore β + γ > β + φ = α.
But since <DCD'>, the angles ⦣BCD and ⦣BCD' are supplements, so
β + γ = 180° - α. Thus β + γ > α implies that 180° - α > α, or
90° > α.   □

Theorem 6

In a Lambert (tri-rectangular) quadrilateral the fourth angle
is acute.

Proof

Let ABCD be a Lambert quadrilateral with right angles at A, B
and C. Let t = L(BC) and let X' = XΓ_t be the reflection of the
plane in line t, with A' = AΓ_t and B' = BΓ_t. The lines L(BA)

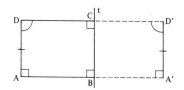

and L(CD) are perpendicular to t and so reflect onto themselves.
The mapping preserves distances and angle measure, so S[AD] ≅
S[A'D'] and ⦣BAD^{o} = 90° = ⦣BA'D'^{o}. Thus AA'D'D is a Saccheri
quadrilateral with congruent, acute summit angles, ⦣ADD' and
⦣A'D'D. Since <DCD'>, ⦣ADC = ⦣ADD', hence ⦣ADC is acute.   □

As a consequence of Th. 4 and Th. 6 , we have the following
result.

Theorem 7

In a Lambert quadrilateral, a side that belongs to an arm
of the acute angle is shorter than the opposite side. (Ex.)

Theorem 7, in conjunction with Th. 2 and Th. 5 gives
us the following information.

Theorem 8

In a Saccheri quadrilateral, the summit is greater than the base and the sides are greater than the altitude. (Ex.)

We know from Th. 5 that the sum of the measures of the angles in a Saccheri quadrilateral is less than $360°$. It will be convenient to use the term "angle sum", in connection with any polygon, to refer to the sum of the measures of its angles. We now want to show that the angle sum of every triangle is less than $180°$. That, of course, will imply that the angle sum of every quadrilateral is less than $360°$.

Theorem 9

The angle sum of a right triangle is less than $180°$.

Proof

Let $\triangle ABC$ be a right triangle, with $\sphericalangle A° = 90°$, $\sphericalangle ABC° = \beta$, and $\sphericalangle ACB° = \gamma$. If M is the midpoint of the hypotenuse S[BC], and F is the foot in L(AB) of M, the fact that $\sphericalangle B$ is acute, (II-3, Cor. 3), implies that F belongs to R(BA). The lines L(CA) and L(MF) are both perpendicular to L(AB), thus are hyperparallel, so all points of L(MF) lie in one side of L(AC). Since M on R(CB) is in the B-side of L(AC), F is in the B-side

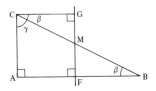

of L(AC), so F ⊂ R(AB) and therefore <AFB>. Now let G denote the foot of C in L(MF). Because $\sphericalangle BMF$ is acute., $\sphericalangle CMF$ is obtuse, so G is not on R[MF) but on the opposite open ray. Thus, $\sphericalangle BMF$ and $\sphericalangle CMG$ are congruent opposite angles. By hypotenuse-angle, $\triangle MBF \cong \triangle MCG$, and so $\sphericalangle MCG° = \sphericalangle MBF° = \beta$ The quadrilateral AFGC

is a Lambert quadrilateral, hence, by Th.6 ∗ACG° < 90°. Since
<FMG>, M ∈ In(∗ACG), and ∗ACM° + ∗MCG° = ∗ACG°. Thus γ + β =
∗ACG° < 90°, and γ + β + ∗A° < 180°.   □

Theorem 10

The angle sum of every triangle is less than 180°.

Proof

If ΔABC is an arbitrary triangle, one of its angles has
maximum measure and we may suppose that ∗C° ≧ ∗A° and ∗C° ≧ ∗B°.
Since a triangle has at most one non-acute angle, (Cor. 3, II-3),
and this must be the largest angle, it follows that both ∗CAB and
∗CBA are acute. Thus if F is the foot of C in L(AB), F must
belong to both R(AB) and to R(BA), and therefore F is between A

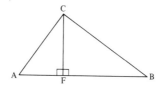

and B. Now the application of Th. 9 to the right triangles, ΔACF
and ΔBCF, yields ∗A° + ∗ACF° + 90° < 180° and ∗B° + ∗BCF° + 90° <
180°, from which it follows that ∗A° + ∗ACF° + ∗BCF° + ∗B° < 180°,
hence that ∗A° + ∗B° + ∗C° is less than 180°.   □

Theorem 10 has the following corollary with a corollary.

Corollary 10.1

The angle sum of every quadrilateral is less than 360°.   (Ex.)

Corollary 10.2

An angle exterior to an S-type quadrilateral at a summit
vertex is greater than the non-adjacent summit angle.   (Ex.)

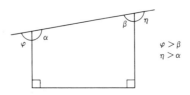

In Euclidean geometry, the angles of a triangle determine
its shape but not its size. In hyperbolic geometry, as the next
theorem indicates, the angles determine both the shape and size.
That is, angle-angle-angle is a new congruence condition and it
implies that in hyperbolic geometry there is no counterpart to the
euclidean theory of similar triangles.

Theorem 11

(angle-angle-angle)   If a correspondence of two triangles
is such that corresponding angles are congruent, then the correspondence is a congruence of the triangles.   (Ex.)

A line through the midpoints of two sides of a triangle is
a "midline" of the triangle. In euclidean geometry, if L(MN)
is a midline of ΔABC bisecting S[CA]and S[CB] at M and N respectively, then L(MN) is parallel to L(AB). Also, because ΔCMN is
similar to ΔCAB, the length of S[MN] is half that of S[AB]. The
triangle similarity does not exist in the present geometry, and it
is instructive to see what analog "triangle midpoint theorem"
does hold.

Theorem 12

The segment joining the midpoints of two sides of a triangle
has length less than half that of the third side. The line of
the segment is hyperparallel to the line of the third side and is
perpendicular to the perpendicular bisector of the third side.

Proof

In ΔABC, let M be the midpoint of S[CA]and N be the midpoint
of S[CB]. Let A*, B*, C* be the feet in L(MN) of A, B, and C
repectively. In ΔCMN, at least one of the angles ⟩CMN and ⟩CNM
is acute, and we may suppose the notation is such that ⟩CNM is acute.
Thus B* is on the open ray opposite to R(NM)and N is between M and B*.
Also ΔBNB* ≅ ΔCNC*, by hypotenuse-angle, hence d(B,B*) = d(C,C*).

The different possibilities correspond to the magnitude of
⚡CMN and to the successive order on the midline of the points A*,
M, C*, N, B*.

Case 1. ⚡CMN° < 90°.

This forces the successive order A*, M, C*, N, B* on L(MN). By
hypotenuse-angle, ΔAMA* ≅ ΔCMC*, and therefore d(A,A*) = d(C,C*).
This equality, with d(B,B*) = d(C,C*), implies that d(A,A*) =
d(B,B*). Thus ABB*A* is a Saccheri quadrilateral with summit

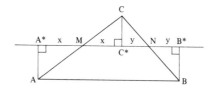

S[AB] and base S[A*B*], so the base line L(MN) is hyperparallel
to the summit line L(AB). If x = d(A*,M*) = d(M,C*) and y = d(C*,N) =
d(N, B*), the order of the points implies that d(M,N) = x + y
and d(A*,B*) = 2x + 2y = 2d(M,N). Since the base is shorter than
the summit, (Th. 8), d(A*,B*) < d(A,B), so 2d(M,N) < d(A,B), hence
d(M,N) < ½d(A,B). The perpendicular bisector of S[AB] is the altitude line of the quadrilateral and hence is perpendicular to the
base line L(MN).

Case 2 ⚡CMN° = 90°.

This equality forces the order A* = M = C*, N, B*. Now d(A,M) =
d(M,C) is also d(A,A*) = d(C*,C), and this equality, with

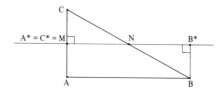

d(C*,C) = d(B*,B), again gives d(A,A*) = d(B,B*). Thus ABB*A* is is again an S-quadrilateral. As in case 1, L(MN) is hyper-

parallel to L(AB) and is perpendicular to the perpendic-

ular bisector of S[AB]. Because $\triangle CA^*N \cong \triangle BB^*N$, $d(A^*N) = d(N,B)$

$= \tfrac{1}{2}d(A^*,B^*)$. Again $d(A^*,B^*) < d(A,B)$ implies that $2\, d(A^*,N) =$

$2d(M,N) <\ d(A,B)$, hence that $d(M,N) < \tfrac{1}{2}d(A,B)$.

Case 3 $\angle CMN^o > 90^o$.   (Ex.)

(Hint: There are three possible point orders, (C^*,M,A^*,N,B^*),

$(C^*,M,A^*=N,B^*)$, and (C^*,M,N,A^*,B^*)),   □

Corollary 12

A line that bisects one side of a triangle and is perpendicu-

lar to the perpendicular bisector of a second side also bisects

the third side.   (Ex).

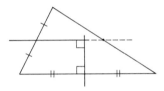

In the last section, we saw that two lines which are perpen-

dicular to a third are hyperparallel. To conclude this section,

we will make use of S-quadrilaterals to establish the converse,

namely that two hyperparallels always have a common perpendicular.

With this property, Th. 5 of the last section gives us a truly

detailed picture of how intersectors and non-intersectors of a line

are related in a pencil.

Theorem 13

If lines r and s are hyperparallel, there exists exactly one

line that is perpendicular to both r and s.

Proof

There cannot be two lines perpendicular to both r and s,

since that would imply the existence of a quadrilateral with four

right angles, contradicting Cor. 10.1. Thus if there is a common
perpendicular to r and s it is unique.

Now let A and B be any two points of r with feet F and G
respectively in s. If either L(AF) or L(BG) is perpendicular to r
then it is a common perpendicular, and we are done. We consider
the case in which neither is perpendicular to r. If d(A,F) =
d(B,G), then AFGB is an S-quadrilateral, and since its altitude line
is perpendicular to both the base and summit lines, we are done.
Finally, we consider the case in which AFGB is an S-type quadrilateral but not an S-quadrilateral. Without loss of generality

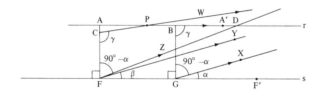

we may suppose that d(A,F) > d(B,G). Let C on R(FA) be the point
such that d(F,C) = d(G,B). Then <FCA>.

At F and G there exist R(FY) and R(GX) parallel to R(AB) and
hence parallel to each other. Let A' and F' be such that <ABA'>
and <FGF'>, and let α = $\angle$F'GX$^\circ$ and β = $\angle$GFY$^\circ$. Because $\angle$F'GX
is an exterior angle to the biangle (X-GF-Y), α > β . Therefore
$\angle$BGX$^\circ$ = 90° - α < $\angle$CFY$^\circ$ = 90° - β . Thus if R(FZ), in the B-G-side
of L(CF), is such that $\angle$CFZ$^\circ$ = 90° - α, then R(FZ) $\subset$ In($\angle$CFY) =
In($\angle$AFY). Therefore R(FY) || R(AB) implies tht R(FZ) intersects
R(AB) at some point D. Now let R(CW) be the ray at C that is parallel to R(FZ). The line L(CW) intersects the side S[AF] of
$\triangle$AFD at C between A and F. By the triangle Pasch property, L(CW)
must intersect a second side. It cannot intersect S[FD], since

L(CW) || L(FD). Thus L(CW) must intersect S(AD) at some point
P between A and D, which implies that {P} = R(CW) ∩ S(AD).

   Because R(GX) || R(AB), and R(AB) is like directed to R(BA'),
R(GX) || R(BA'). The biangles (A'-BG-X) and (W-CF-Z) have congruent
segment sides and congruent angles at G and F. Thus, by segment-
angle, the biangles are congruent, hence γ = ∦A'BG° = ∦PCF°.

   On R(BA'), let Q be the point such that d(B,Q) = d(C,P),
and let P* and Q* denote the feet of P and Q respectively in s.

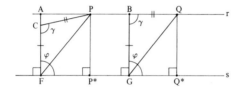

By side-angle-side, ΔFCP ≅ ΔGBQ, hence S[PF] ≅ S[QG], and ∦CFP°
= ∦BGQ° = φ. Therefore ∦PFP*° = 90° - φ = ∦QGQ*°. Now ΔPFP* ≅
ΔQGQ*, by hypotenuse-angle, so S[PP*] ≅ S[QQ*]. Thus P*Q*QP is
a Saccheri quadrilateral, and its altitude line is perpendicular
to the summit line L(PQ) = r and to the base line L(P*Q*) = s.   □

   In Th. 4 III-2, it was shown that if P is not on line t, then
to each point Y in t there corresponds a line s in P(P) which is
hyperparallel to t, and s and t form congruent alternate interior
angles with the transversal L(PY). It is easily shown that if Y_1
and Y_2 are distinct points of t then they determine different
hyperparallels in P(P), otherwise the angle sum of ΔPY₁Y₂ would
be 180°. With Th. 12, we can now establish a converse property,
namely that each hyperparallel to t in P(P) is a hyperparallel
corresponding to some point Y in t under the correspondence of
Th. 5, III-2.

Consider two lines r and t perpendicular to a line u at
points P and F respectively. Let line s in the pencil P(P)
be a hyperparallel to t that is not r, and let R(PZ) be the ray
on s such that ⧊FPZ is acute. By Th.12, there exists a line v
which is perpendicular to s at some point H and to t at some point
G. Because u ⫝̸ s, v ≠ u, and since u and v are both perpendicular to t, u)(v. Thus H and G on v are in one side of u. Because the quadrilateral FGHP has right angles at F, G and H,
⧊FPH is acute, which implies the R(PH) = R(PZ), and so v is in
the Z-side of u.

Corresponding to M, the midpoint of S[GH], there is a point
reflection Γ_M that interchanges G and H and leaves v invariant.

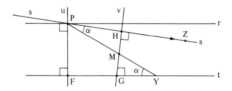

Thus s, perpendicular to v at H, must map to t, perpendicular
to v at G. If Y = $P\Gamma_M$, then P ∈ s implies that Y ∈ t. Because
Γ_M interchanges the sides of L(PY), L(PY) separates R(PH) and
R(YG), and since Γ_M preserves angle measure, ⧊YPH° ⁼ ⧊PYG° = α.
Thus s corresponds to Y in the correspondence of Th.4, III-2.
That correspondence is therefore a 1-1 correspondence between the
intersectors of t in P(P) and the hyperparallels to t in P(P).

Let R(PC) be the ray on r in the Z-side of u. One arm
of the fan angle of P and t intersects S(FC) at a point D, and
R(PD) ⊂ In(⧊YPZ). As d(F,Y) increases, α decreases, and it is
not difficult to show that both ⧊DPY and ⧊DPZ decrease. In descriptive terms, as Y moves away from F, in one side of u,

the rays R(PY) and R(PZ), on the corresponding intersector and
hyperparallel respectively, move toward each other from opposite

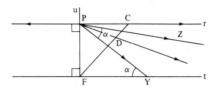

sides of the parallel L(PD). The same phenomenon occurs in the
opposite side of u.

Exercises - Section 3

1.   Prove Th. 3.

2.   Prove Th. 4.

3.   Prove Th. 7

4.   Prove Th. 8

5.   Prove Cor. 10.1

6.   Prove Cor, 10.2

7.   Prove Th. 11.

8.   Complete Case 3 in the proof of Th. 12.

9.   Prove Cor. 12.

10.   In the proof for Th. 13, show that Q cannot be P.

11.   If lines r = L(PZ) and s = L(Y$_1$Y$_2$) are hyperparallel,
      show that L(PY$_1$) and L(PY$_2$) as transversals of r and s
      cannot both form congruent alternate interior angles
      with r and s.

12.   If r ⊥ L(AB) at A and s ⊥ L(AB) at B, and M is the mid-
      point of S[AB], prove that a transversal t of r and s

forms congruent alternate interior angles with r and s
if and only if M ∈ t.

13. A quadrilateral whose four sides have the same length and
    whose four angles have the same measure is sometimes called
    a "pseudosquare". Show that pseudosquares exist. Why
    must the lines of opposite sides of a pseudosquare be hyper-
    parallel?

14. If the vertices of a quadrilateral ABCD belong to a circle
    C(O,r), then in euclidean geometry the diagonally opposite
    angles at A and C are supplements, as are those at B and D.
    Why is this not possible in hyperbolic geometry? Show
    that in absolute geometry (hence in both hyperbolic and
    and euclidean geometry) the sum of the measures of one
    diagonally opposite pair equals the sum of the measures
    of the other diagonally opposite pair.

15. Prove that if all triangles in absolute geometry have
    the same angle sum k, then k must be 180°.

16. A Saccheri quadrilateral ABCD has base S[AB] and altitude
    S[MN], with M the midpoint of S[AB]. (a) Prove that the
    diagonals of S[AC] and S[BD] intersect at a point E on S[MN].
    (b) If P and Q are the midpoints of the sides, show that
    P and Q have a common foot F in L(MN). (c) Show that
    F ≠ E and explain which side of L(PQ) contains E. (d)
    If O is the midpoint of S[MN], where is O in relation to
    E and F?

Section 4. Angle of Parallelism Function, Triangle Defect, Distance
Variations

Our primary goal in this section is to investigate the manner
in which the distance from a point P to a line t changes as P varies
on a line s. However, to help in such a study we need another concept which is of great importance in its own right, namely the
angle of parallelism function. We will therefore begin by establishing the nature of this function, and then turn to its application in distance variation problems.

Definition

(Angle of parallelism function)   The angle of parallelism
function π has the function value π(x) equal to the measure of the
acute angle in a right biangle whose segment side has length x.

If $0 < x < \infty$, there exists a right biangle whose segment side
has length x. All such biangles are congruent, by segment-angle,
so π is a well defined function for $0 < x < \infty$.

Theorem 1.

The function π is strictly decreasing, $(y > x \Rightarrow \pi(y) < \pi(x))$.

Proof

Let P, Q be points on a ray R(AB) such that $d(A,P) = x$ and
$d(A,Q) = y$, with $y > x$. There exists a line L(AC) which is perpendicular to L(AB) at A. At P and Q there exist rays R(PX) and R(QY)
respectively such that R(PX) || R(AC) and R(QY) || R(AC). Because
R(PX) || R(QY), the biangle (X-PQ-Y) exists. Since $y > x$, then
<APQ>, and this implies that ⊁APX is an exterior angle at P to
(X-PQ-Y). By the exterior angle property of biangles, (Th.6, III-2),
⊁PQY° < ⊁APX°, hence ⊁APY° < ⊁APX°, which is to say that $\pi(y) < \pi(x)$ □

Since a strictly decreasing function is a one-to-one mapping
of the domain onto the range, Th. 1 has the following corollary.

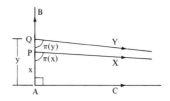

Corollary 1

The inverse relation to π is a function π⁻¹

Since the function value $\pi(x)$ is the measure of an acute angle, it is clear that the range of π is contained in the interval $0 < y < 90$. However, it is neither obvious to see, nor easy to prove that the range of π is the whole open interval. To settle the question, we need the help of a triangle property that we have not yet established and we digress, temporarily, to obtain this property.

Definition

(Triangle defect)   The defect of a triangle is the amount by which its angle sum is less than 180°. Thus the defect of $\triangle ABC$, denoted by $D(ABC)$, is the number $180° - \angle A° - \angle B° - \angle C°$.

Definition

(Simple transversal; subtriangles).   A simple transversal of a triangle is a line that subdivides an angle of the triangle. If line t is a simple transversal of $\triangle ABC$ and subdivides $\angle A$ it intersects $S(BC)$ as some point D, and the triangles $\triangle ADB$ and $\triangle ADC$ are the subtriangles corresponding to t.

The next theorem states the triangle property we seek.

Theorem 2

The defect of a triangle is the sum of the defects of the two subtriangles that correspond to a simple transversal of the triangle.

Proof

Let t denote a line that subdivides $\angle BAC$ of $\triangle ABC$ and which intersects $S(BC)$ at a point D. Let $\alpha_1 = \angle DAB°$, $\alpha_2 = \angle DAC°$,

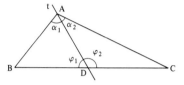

$\varphi_1 = \angle ADB°$, and $\varphi_2 = \angle ADC°$. By definition,

$$D(ADB) = 180° - \alpha_1 - \varphi_1 - \angle B°, \qquad (1)$$

and

$$D(ADC) = 180 - \alpha_2 - \varphi_2 - \angle C° \qquad (2)$$

Because $<BDC>$, $\alpha_1 + \alpha_2 = \angle A°$ and $\varphi_1 + \varphi_2 = 180°$. With these equalities, (1) and (2) imply that

$$D(ADB) + D(ADC) = 360° - (\alpha_1 + \alpha_2) - (\varphi_1 + \varphi_2) - (\angle B° + \angle C°)$$
$$= 180° - (\angle A° + \angle B° + \angle C°) = D(ABC). \qquad \square$$

Corollary 2

The defect of a triangle is greater than the defect of either of the subtriangles that correspond to a simple transversal of the triangle.

As an interesting aside, in euclidean geometry the area of $\triangle ABC$, in the proof of Th.2, would of course be the sum of the areas of the subtriangles $\triangle ADB$ and $\triangle ADC$. Thus Th.2 suggests that in hyperbolic geometry the area of a triangle might be proportional to its defect. This suggestion is correct, though its proof and indeed the consideration of area is outside the scope of this book.

Returning to the function π, we want to prove that if α is a number between 0 and 90, then there exists a positive number a such that $\pi(a) = \alpha$. This is equivalent to the claim that if $\angle BAC$ is any acute angle, then $\angle BAC$ is also the acute angle of some

right biangle. If the angle arm R[AC) is a ray side of such a biangle
then the line of the other ray side must be perpendicular to L(AB),
(at a point of R(AB)), and must also be parallel to R(AC). Let us
see, first, if we can establish the existence of some line that is
perpendicular to R(AB) and does not intersect L(AC).

Let h be any positive number, and on the arm R[AB) of the acute
angle $\angle$BAC let A_1 be such that $d(A,A_1) = h$. Let t_0 and $_1$ designate
the lines that are perpendicular to L(AB) at A and A_1 respectively.

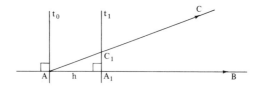

Now, $t_1) (t$, so $t_1 \subset$ C-side of t_0. Thus if t_1 intersects L(AC) it
intersects R(AC). If $t_1 \cap R(AC) = \emptyset$, then t_1 is a line that is
perpendicular to R(AB) and does not intersect L(AC). Suppose,
instead, that t_1 intersects R(AC) at some point C_1. Now we consider
the point A_2 on R(AB) such that $d(A,A_2) = 2h$ and the line $t_2 \perp L(AB)$
at A_2. As before, if $t_2 \cap R(AC) = \emptyset$, then t_2 is a line we are seeking,
namely a line perpendicular to R(AB) that does not intersect L(AC).
Thus we again suppose that t_2 intersects R(AC) at some point C_2. But
now, before going on to a point A_3 and a line t_3, we bring triangle
defects into the picture. On $R(A_2C_2)$ let D_2 be the point such that

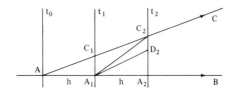

$d(A_2,D_2) = d(A_1,C_1)$. If $\delta = D(A_1A\ C_1)$, then from $\Delta AA_1C_1 \cong \Delta A_1A_2D_2$
it follows that $\delta = D(A_1A_2D_2)$. Because t_o, t_1, and t_2 are hyperparallels, and $\langle AA_1A_2 \rangle$, t_1 separates t_0 and t_2 and therefore separates
A and C_2. From $\langle AC_1C_2 \rangle$, it follows that $\not{\star}A_1C_1C_2$ is the supplement of
the acute angle $\not{\star}A_1C_1A$ and is therefore obtuse. Thus, in the S-type
quadrilateral $A_1C_1C_2A_2$, $\not{\star}A_1C_1C_2$ is the larger summit angle and therefore $d(A_2,C_2) > d(A_1,C_1)$, which implies that $\langle A_2D_2C_2 \rangle$.

Now consider the defect of ΔAA_2C_2. Because $L(A_1C_2)$ is a simple
transversal of the triangle, Th. 2 implies that

$$D(AA_2C_2) = D(AA_1C_2) + D(A_2A_1C_2). \qquad (1)$$

Since $L(A_1C_1)$ is a simple transversal of ΔAA_1C_2, Cor. 2 implies that

$$D(AA_1C_2) > D(AA_1C_1) = \delta. \qquad (2)$$

Also, since $L(A_1D_2)$ is a simple transversal of $\Delta A_2A_1C_2$, Cor. 2 implies
that

$$D(A_2A_1C_2) > D(A_1A_2D_2) = \delta. \qquad (3)$$

Applying (2) and (3) in (1), we obtain

$$D(AA_2C_2) > 2\delta. \qquad (4)$$

Starting with the triangle ΔAA_1C_1, with defect δ, the intersection of t_2 with $R(AC)$ led us to a triangle ΔAA_2C_2 with a defect greater than 2δ. Now the process can be repeated. Let A_3 on $R(AB)$ be such
that $d(A,A_3) = 3h$ and let t_3 be the line perpendicular to $L(AB)$ at
A_3. If $t_3 \cap R(AC) = \emptyset$, we are done. If $t_3 \cap R(AC)$ is a point C_3,
then there exists a point D_3 on $R(A_3C_3)$ such that $d(A_3,D_3) = d(A_1,C_1)$.
As before, $\langle A_3D_3C_3 \rangle$ and $D(A_2A_3D_3) = \delta$.

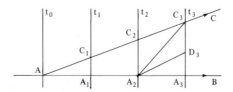

Because $L(A_2C_3)$ is a simple transversal of ΔAA_3C_3, it follows from
Th. 2 that

$$D(AA_3C_3) = D(AA_2C_3) + D(A_3A_2C_3). \qquad (5)$$

Since $L(A_2C_2)$ is a simple transversal of ΔAA_2C_3, Cor. 2 and (4) imply that

$$D(AA_2C_3) > D(AA_2C_2) > 2 \delta. \tag{6}$$

Since $L(A_2D_3)$ is a simple transversal of $\Delta A_3A_2C_3$, Cor. 2 implies that

$$D(A_3A_2C_3) > D(A_2A_3D_3) = \delta. \tag{7}$$

Applying (6) and (7) in (5), we obtain

$$D(AA_3C_3) > 3\delta. \tag{8}$$

The positive integers are unbounded above, so there exists a positive integer m such that

$$m > (180°)/\delta. \tag{9}$$

On $R(AB)$ there exist successive points $A_1, A_2, \ldots, A_m$, at distances h, 2h, $\ldots$, mh from A, and there exists a line $t_i \perp L(AB)$ at A_i, i = 1, 2, $\ldots$, m. Assume that all of the lines t_1, t_2, $\ldots$, t_m intersects $L(AC)$. Then, by the constructive process described, there must exist a triangle ΔAA_mC_m such that

$$D(AA_mC_m) \geq m \delta. \tag{10}$$

By the definition of triangle defect, (10) and (9) imply that

$$180° - (\sphericalangle A° + \sphericalangle A_m° + \sphericalangle C_m°) \geq m \delta > \frac{180°}{\delta} \cdot \delta = 180°,$$

hence that

$$\sphericalangle A° + \sphericalangle A_m° + \sphericalangle C_m° < 0. \tag{11}$$

But no triangle has a negative angle sum, since angle measures are positive numbers, thus the inequality (11) is impossible. Therefore the assumption that all the lines t_1, t_2, $\ldots$, t_m intersect $L(AC)$ cannot hold, and at least one of the lines must be a non-intersector of $L(AC)$. We state this result, for reference, as a lemma.

Lemma

If $\sphericalangle BAC$ is acute, there exists a line which does not intersect $L(AC)$ and is perpendicular to $L(AB)$ at a point of $R(AB)$.

With the lemma we now have all that we need to prove that the range of π is the whole open interval (0,90). It will simplify the proof it we make use of the following concept, which is also meaningful in absolute geometry.

Definition

(Projection into a line)   The projection of set s into a line t is the mapping f in which X ∈ s and X' = f(X) is the foot in t of point X. Point X projects to the image point X' and s' = f(s) is the projection of s in t.

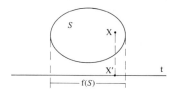

Theorem 3

If α is a number between 0 and 90, there exists a positive number a such that π(a) = α.

Proof

Let ∡BAC be an acute angle of measure α. At each point X of R[AB] there is a line perpendicular to L(AB) and we denote this perpendicular by t_X. Each two of these lines are hyperparallel and hence non-intersecting. By the lemma, there exists D ∈ R(AB) such that t_D ∩ L(AC) = ∅. If P is a point of R(AC) with foot P' in L(AB), then P' ∈ R(AB) = R(AD) because ∡BAC is acute. Since $t_{P'}$)(t_D, all points of $t_{P'}$ are in the P-side of t_D, which is also the A-side of t_D. Therefore P' belongs to R(DA). From P' ∈ R(AD) ∩ R(DA) it follows that P' ∈ S(AD). Thus the whole open ray R(AC) projects into a subset of S(AD).

Our plan now is to find the point X on S[AD] nearest to A and such that t_X ∩ L(AC) = ∅ and show that this line is also paral- lel to R(AC). Toward this end, we define sets s_1 and s_2 as follows:

$$s_1 = \{X:\ X ∈ S[AD]\ \text{and}\ t_X ∩ L(AC) ≠ ∅\};$$
$$s_2 = \{X:\ X ∈ S[AD]\ \text{and}\ t_X ∩ L(AC) = ∅\}.$$

We will show that s_1, s_2 form a Dedekind cut of S[AD]. Clearly, the

definitions of s_1 and s_2 imply that $S[AD] = s_1 \cup s_2$, that $A \in s_1$,
$D \in s_2$, and that $s_1 \cap s_2 = \emptyset$. Finally, consider $\langle AYX \rangle$ and $X \in s_1$.
Since $X \in s_1$, t_X intersects $L(AC)$ at a point X' which obviously
belongs to $R(AC)$. By the Pasch property of $\triangle AXX'$, because t_Y
intersects side $S[AX]$ it must intersect a second side. But
$t_Y \cap t_X = \emptyset$ implies that $t_Y \cap S[XX'] = \emptyset$. Therefore t_Y must intersect $S[AX']$, hence must intersect $L(AC)$, so $Y \in s_1$. Thus s_1 and
s_2 satisfy the requirements for a Dedekind cut of $S[AD]$, and there
exists a Dedekind point E on $S[AD]$ such that $\langle AXE \rangle \Rightarrow X \in s_1$
and $\langle EXD \rangle \Rightarrow X \in s_2$. Now, we want to show that t_E is the line we
are looking for, namely that t_E is parallel to $R(AC)$.

Because $R(AC)$ projects into $S(AD)$, and the projection points
belong to s_1, $E \neq A$. Assume that $E \in s_1$. Then t_E intersects $R(AC)$
at some point E_o. Let P on $R(AC)$ be any point such that $\langle AE_o P \rangle$
and let P' be the foot of P in $L(AB)$ and hence a point of $S(AD)$.
Because $t_E \,)(\, t_{P'}$, all points of $t_{P'}$ lie in one side of t_E. Since

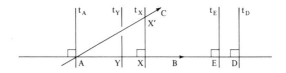

$\langle AE_o P \rangle$, t_E separates A and P therefore separates A and $t_{P'}$, so
E is between A and P'. But $\langle AEP' \rangle$ with $P' \in S(AD)$ implies that
$\langle EP'D \rangle$, hence that $P' \in s_2$. Therefore $t_{P'} \cap L(AC) = \emptyset$. But $P \in R(AC) \cap t_P$
implies that $t_{P'} \cap L(AC) \neq \emptyset$. The contradiction shows that $E \in s_1$
is impossible, hence that $E \in s_2$. We have shown that $t_E \cap L(AC) = \emptyset$.
It remains to be shown that $t_E \parallel R(AC)$.

On t_E there is a ray $R(EZ)$ in the C-side of $L(AB)$. We want to show that $R(EX) \subseteq In(\measuredangle AEZ)$ implies that $R(EX) \cap R(AC) \neq \emptyset$.

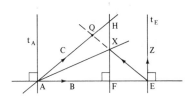

If X is on $L(AC)$, or in the non-E-side of $L(AC)$, then clearly $R(EX)$
intersects $L(AC)$. Because Z, X, and C are all in one side of $L(AB)$,
the ray $R(EX)$ intersects $R(AC)$. Assume that X is in the E-side
of $L(AC)$. and let F be its foot in $L(AB)$. Because $R(EX)$ is
interior to the right angle $\measuredangle AEZ$, $\measuredangle AEX$ is acute, hence $F \in R(EA)$.
From $X \in Z$-C-side of $L(AB)$ and $X \in$ E-B-side of $L(AC)$ it follows
that $X \in In(\measuredangle BAC)$. Therefore $\measuredangle BAX^\circ < \measuredangle BAC^\circ$, so $\measuredangle BAX$ is acute,
which implies that $F \in R(AB) = R(AE)$. Because $F \in R(EA) \cap R(AE)$,
<AFE>, and so $F \in s_1$ and therefore t_F intersects $R(AC)$ at some
point H. Since $L(EX)$ intersects side $S[FH]$ of $\triangle AFH$, it must intersect a second side. Since $L(EX)$ intersects $L(AB)$ at E, and E is
not on $S[AF]$, $L(EX) \cap S[AF] = \emptyset$. Therefore $L(EX)$ must intersect
$S[AH]$ at some point Q. Since $Q \neq A$, $Q \in S(AH] \subseteq R(AC)$, and so
$L(EX)$ intersects $R(AC)$. Because Q is in the C-side of $L(AB)$, it
is not on the closed ray opposite to $R(EX)$, and therefore $Q \in R(EX)$
$\cap R(AC)$. In all cases, then, $R(EX) \subseteq In(\measuredangle AEZ)$ implies that $R(EX) \cap$
$R(AC) \neq \emptyset$, and therefore $R(EZ) \parallel R(AC)$.

Now define $a = d(A,E)$. Because (C-AE-Z) is a right biangle
whose acute angle $\measuredangle EAC = \measuredangle BAC$ has measure α, $\pi(a) = \alpha$.   □

Corollary 3.1

If $\measuredangle BAC = \alpha < 90^\circ$, and if point E on $R(AB)$ is such that
$d(A,E) = \pi^{-1}(\alpha)$, then the line perpendicular to $L(AB)$ at E is
parallel to $R(AC)$.

Corollary 3.2

If (C-AE-Z) is a right biangle, with acute angle at A. the projection of R(AC) into line L(AE) is the open segment S(AE).

We turn now to the distance variation problem mentioned at the start of the section. If point P in line s has foot F in line t, we want to know the manner in which $d(P,F) = d(P,t)$ changes as P varies on s. We begin with the case in which s and t intersect at a point A and we define $x = d(A,P)$, $y = d(P,F)$ and $z = d(A,F)$.

If $s \perp t$, $F = A$, $y = x$ and $z = 0$. As x increases without bound, so does y, and z is always zero. If $s \not\perp t$, let ⊀BAC be an acute angle, whose measure is α, formed by s and t, with $R(AC) \subset s$. Consider two points P and P_1 on R(AC), such that $<APP_1>$ and let F and F_1 be their respective feet in t. Because ⊀BAC is acute, F and F_1 belong to R(AB), and $<APP_1>$, which implies that $<AFF_1>$. The figure PFF_1P_1 is an S-type quadrilateral.

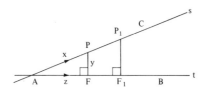

Because $<APP_1>$, the angle ⊀FPP_1 is the supplement of the acute angle ⊀FPA, hence is obtuse, and hence is the larger of the two summit angles in the quadrilateral. Therefore $d(P_1,F_1) > d(P,F)$. Thus as x increases y increases, and since $<AFF_1>$ implies that $d(A,F_1) > d(A,F)$, z also increases.

Let E on R(AB) be the point whose distance from A is $\pi^{-1}(\alpha)$ and let v denote the line perpendicular to t at E. By Cor. 3.1, v || R(AC). If R(EW) on v lies in the C-side of t, then (C-AE-W) is a right biangle and, by Cor. 3.2, the projection of R(AC) into

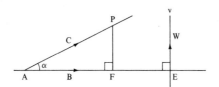

t is the open segment S(AE). Thus z = d(A,F) is always less

than d(A,E) = $\pi^{-1}(\alpha)$. Since z + y > x, z + y must become unboun-

dedly large as x increases without bound. But since z is less than

$\pi^{-1}(\alpha)$, y must become unboundedly large. We summarize these results

as a theorem.

<u>Theorem 4</u>

If ⊀BAC is acute, a point P on R(AC) has foot F in L(AB) on

the ray R(AB). If x = d(A,P), y = d(P,F) and z = d(A,F), then

y and z are increasing functi̲s of x. As x increases without bound,

so does y, but z is always less than $\pi^{-1}(⊀BAC°)$.

Next, let us consider the case in which s is parallel to t.

Let A be any point of s, with foot B in t, and let R(AX) in s be

parallel to R(BY) in t. Consider two points P and P_1 in R(AX),

with feet F and F_1 respectively in t, and such that <APP$_1$>.

Because the lines L(AB), L(PF), and L(P$_1$F$_1$) are all perpendicular

to t, each two are hyperparallel. Since <APP$_1$>, L(PF) separates A

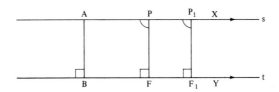

and P_1, and therefore separates L(AB) and L(P$_1$F$_1$), which implies

that <BFF$_1$>. The angle ⊀APF is an exterior angle at P to the biangle

$(P_1\text{-}PF\text{-}F_1)$, hence is greater than the right angle $\angle PFF_1$. Because
it is obtuse, $\angle APF$ is the larger of the summit angles in the
S-type quadrilateral ABFP, and therefore $d(A,B) > d(P,F)$. If X
and Y are chosen so that $<PP_1X>$ and $<FF_1Y>$, then $\angle PP_1F_1$ is an
exterior angle at P_1 to $(X\text{-}P_1F_1\text{-}Y)$. Thus $\angle PP_1F_1$ is greater than
the right angle $\angle P_1F_1Y$, and is therefore the larger of the summit
angles in the S-type quadrilateral PFF_1P_1. Thus, $d(P,F) > d(P_1,F_1)$.

If we set $s = d(A,P)$, $y = d(P,F) = d(P,t)$, and $z = d(B,F)$,
the argument above shows that as x increases, i.e. as P varies
in the direction of parallelism, y decreases. Because $<APP_1>$
implies $<BFF_1>$, and hence $d(B,F) < d(B,F_1)$, as x increases so does z.
Moreover, corresponding to any positive number h, there exists a
point F' on R(BY) such that $d(B,F') = h$. The line perpendicular
to t at F' is hyperparallel to L(AB) and cannot intersect S[AB].
By the Pasch properties of the biangle (X-AB-Y), (Th. 1, III-2),
the line must intersect a second side of the biangle and hence
must intersect R(AX) at some point P'. Thus as x increases without
bound, so does z.

In the situation above, it is still not clear how small
y can become as x gets large. Also, if $x = d(A,P)$, but P is on
the ray opposite to R(AX), then clearly y increases with x. But
how large can y become? The next theorem answers these questions
and also summarizes the results of the preceding discussion.

Theorem 5

If line s is parallel to line t, the distance $d(P,t)$ decreases
at P varies in the direction of parallelism on s and increases as
P varies on s in the opposite direction. Moreover, if k is any
positive number, there is exactly one point P_0 on s such that
$d(P_0,t) = k$.

Proof

The decreasing and increasing variation of d(P,t) has already been established and obviously implies that there can be at most one point P_0 on s such that $d(P_0,t) = k$. We still need to show that such a point P_0 exists.

Now let A be any point in s, with foot B in t, and let R(AX) in s be parallel to R(BY) in t. If d(A,B) = k, then A is the desired point P_0, so we suppose that $d(A,B) \neq k$. Then d(A,B) is either smaller or larger than k and we consider the two possibilities separately.

Case 1

d(A,B) < k. On R(BA) there is a point C such that d(B,C) = k, and d(B,A) < d(B,C) implies that <BAC>. Let W on t be such that <WBY>. At C there is a ray R(CZ) || R(BW). The line s intersects side S[BC] of the biangle (W-BC-Z) at A between B and C. Since s is not parallel to R(BW), the Pasch property of the biangle implies that s must intersect R[CZ] or R[BW]. But s || t implies that s ∩ R[BW] = ∅, so s must intersect R[CZ) at some point D which obviously must belong to R(CZ), and we may suppose that <CDZ>.

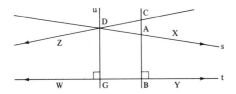

Let G be the foot of D in t and set u = L(DG). Because u and L(AB) are hyperparallel, and D on u is in the Z-W-side of L(AB), then G on u is in the Z-W-side of L(AB), and we may suppose that <WGB>. Since L(AB) separates D and X it also separates G and Y, and <DAX> and <GBY> imply that R(DA) || R(GB), (since

R(AX)||R(BY)). Also, <CDZ> and <BGW> and R(CZ) || R(BW) imply
that R(DZ) || R(GW). Thus ∢ADZ is the fan angle of D and t.
Since u passes through D and is perpendicular to t, it bisects
∢ADZ. Setting v = L(DZ), it follows that the reflection of the
plane in u must interchange v and s. Thus C on v is mapped by
Γ_u to a point C' = $C\Gamma_u$ on s. Because t ⊥ u, $t\Gamma_u$ = t, and so B on
t maps to B' on t. Since Γ_u is a motion of the plane, d(C,B) =
d(C',B') = k. Therefore C' = P_o is the desired point on s such
that d(P_o,t) = k.

Case 2

    d(A,B) > k.   (Ex.)   □

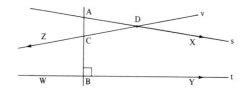

Corollary 5

    If s || t, as P varies on s in the direction of parallelism,
d(P,t) becomes arbitrarily small, and as P varies on s in the
opposite direction, d(P,t) becomes arbitrarily large.

    As Cor. 5 indicates, if s || t then s and t approach each
other in the directions of parallelism arbitrarily closely without
ever touching. For this reason, parallel lines s and t are called
"asymptotes" in many texts. A drawing that illustrates the approach of
s and t in the direction of parallelism makes the lines appear

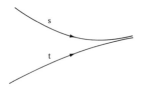

'curved'. But a diagram, of course, is not part of any mathematical argument. It is simply a visual aid in following the steps
in an argument, or a visual suggestion of some property. Thus the
appearance of a diagram, or in fact its removal, does not affect
the reasoning in any proof. The 'straightness' of a line s, for
example, is represented mathematically by the property that if
A,B,C are successive points on s, the number d(A,C) is the number
d(A,B) + d(B,C), and this is independent of how s may appear in
a drawing.

Our final case is that in which s and t are hyperparallel.
Let u denote the line that is perpendicular to s at A and to t at B
and let X on s and Y on t be in one side of u. Consider points P

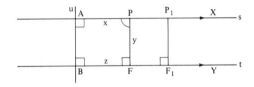

and P_1 on R(AX), with feet F and F_1 respectively in t, and such
that <APP_1>. Because ABFP is an L-quadrilateral, with acute
angle at P, d(A,B) < d(P,F). Since <APP_1>, $\nleq FPP_1$ is the supplement of the acute angle $\nleq APF$, hence is obtuse, hence is the larger
of the summit angles in the S-type quadrilateral PFF_1P_1. Therefore
d(P,F) < d(P_1,F_1). Since <APP_1> clearly implies <BFF_1>, d(B,F) <
d(B,F_1). Letting x = d(A,P), y = d(P,F), and z = d(B,F), it
follows that y and z are increasing functions of x.

To show that y increases without bound as x does, let h be
any positive number. At B there is a ray R(BZ) || R(AX). Since
$\nleq$ABX is the acute angle of the right biangle (X-AB-Z),
R(BZ) $\subset$ In($\nleq$ABY), and so $\nleq$YBZ is acute. If Q is variable on R(BZ),
and r = d(B,Z), it follows from Th.4 that as r increases

without bound, so does d[Q,L(BY)] = d(Q,t)   Thus there exists a

value r = r_0 , and a corresponding point Q_0, such that d(Q ,t) > h.

Let F_0 in R(BY) be the foot in t of Q_0.   The line L(F_0Q_0) is

hyperparallel to L(AB), since both are perpendicular to t, so

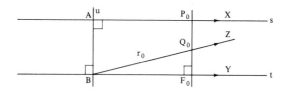

L(F_0Q_0) ∩ S[AB] = ∅.   Since L(F_0Q_0) intersects the ray side R[BZ]

of the biangle (X-AB-Z), and does not intersect S[AB] , it must

intersect R(AX) at some point P_0.   Because Q_0 ∈ In(∢ABY),

Q_0 ∈ A-side of L(BY) = s-side of t.   Since L(BZ) || s, all points

of L(BZ) are in the B-side of s, which is the t-side of s.   Thus,

Q_0 ∈ (s-side of t) ∩ (t-side of s), so Q_0 is between s and t.   By

Th. 5, II-4, it follows that Q_0 is between P_0 and F_0 .   Therefore

d(P_0,F_0) > d(Q_0,F_0) > h.   Thus as x increases without bound, y = d(P,t)

becomes larger than any preassigned number h, so y increases with-

out bound.

As x increases, z = d(B,F) also increases, but not without

bound.   If β = ∢YBZ°, there exists E on R(B,Y) such that d(B,E) =

π^{-1}(β).   By Cor. 3.1, the line v that is perpendicular to t at E

is parallel to R(BZ).   Since v and s are both parallel to R(BZ),

they are parallel to each other. (Th. 12, III-1).

Because s || v, s lies in one side of v, hence s ⊂ A-B-side of v.

In particular, R(AX) is also contained in the X-Y side of u.

Thus R(AX) lies in the strip between the hyperparallels v and u.
Therefore the projection of R(AX) into t is contained in S(BE).
But, by the argument of the previous paragraph, a perpendicular to
t at a point of S(BE) must intersect R(AX). Thus the projection of
R(AX) into t is the open segment S(BE). Thus z = d(B,F), is
bounded by d(B,E) = $\pi^{-1}(\beta)$.

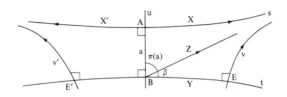

    Under reflection of the plane in u, the lines s and t are
invariant because both are perpendicular to u. The ray R(AX)
maps to the opposite open ray R(AX'). Line v, which is perpendicular to t at E and parallel to R(AX), maps to v' perpendicular to
t at E' and parallel to R(AX'). The projection of R(AX) into t
is S(BE), and the projection of R(AX') iinto t is S(BE'). Thus
we have the interesting phenomenon that the projection of the whole
line s into t is the open segment S(EE'). We note also that while
s is between v and v', (since s is contained in the open strip
St(vv'), s does not separate v and v' because both lines are in
the t-side of s. We have here, too, our first instance of three
lines, s, v, and v' with the property that no line is a transversal
of all three of them. The proof is left as an exercise.
    We summarize our findings in the following theorem.

Theorem 6

    If lines s and t are perpendicular to line u at points A and
B respectively, with d(A,B) = a > 0, and if P in s is not A, then

P and its foot F in t lie in one side of u. If $x = d(A,P)$,
$y = d(P,F)$, and $z = d(A,F)$, then y and z are increasing functions
of x. The minimum value of y is a, achieved at $x = 0$. As x
increases without bound, so does y, but z is bounded above by
$\pi^{-1}[90^{\circ}- \pi(a)]$. If E and E' are the two points of t at distance
$\pi^{-1}[90^{\circ}- \pi(a)]$ from B the projection of s into t is the open segment S(EE').

Exercises - Section 4

1. Prove Case 2 for Th. 5.

2. Given two lines r and s, how many lines which are perpendicular to r are also parallel to s? Explain your
answer in each of the following cases: **(i)** r $\perp$ s; **(ii)**
r $\not\perp$ s and r $\cap$ s $\neq \emptyset$; (iii) r $||$ s; (iv) r)(s.

3. A quadrilateral ABCD has a circumcircle (a circle passing
through all four vertices) if and only if the four perpen-
cicular bisectors of the sides belong to a pencil. If
ABCD is a Saccheri quadrilateral with base S[AB] of
fixed length 2a, but with variable side length $x = d(A,D)$
$= d(B,C)$, explain the existence of a number m such that
the quadrilateral has a circumcircle if and only if
$x < m$.

4. The defect of a quadrialteral is defined to be the
amount by which its angle sum is less than 360°. If
line t bisects the sides S[AC] and S[BC] of $\triangle$ABC, and
if A has foot A* in t and B has foot B* in t, prove that
the defect of the triangle $\triangle$ABC equals the defect of the
S-quadrilateral ABB*A*.

5.   Show that lines lines L(AB), L(CD), and L(EF) exist which
     form a configuration that might be called an "infinite
     triangle" in the sense that R(AB) || R(CD), R(DC) || R(EF),
     and R(FE) || R(BA).

6.   Show that no line is a transversal of all three lines
     L(AB), L(CD) and L(EF) in Ex. 5.

7.   Lines v and v' are perpendicular to t at E and E' respec-
     tively, and line u is perpendicular to t at B, the midpoint
     of S[EE']. If $\beta = \pi[d(E,B)]$, and $\alpha = 90° - \beta$, there
     exists a point A on u such that $d(B,A) = \pi^{-1}(\alpha)$. If
     line s $\perp$ u at A, show that no line is a transversal of
     s, v and v'.

Section 5. The 3-Point Property, Cycles

It is axiomatic in absolute geometry, hence in both euclidean and hyperbolic geometry, that two points belong to a unique line. In euclidean geometry, three non-collinear points have the "3-point property" that they belong to a unique circle. Thus every euclidean triangle has a circumcircle. In this section we wish to investigate the counterpart of the 3-point property in hyperbolic geometry.

It will expedite the discussion to introduce names for two special types of line collections.

Definition

(Parallel family)   The <u>parallel family in the direction</u> of R(AB) is the collection of lines, denoted by F [R(AB)] , which consists of the line of R(AB) and all lines which are parallel to R(AB).

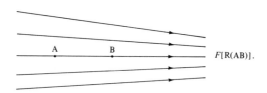

$F[R(AB)]$.

There is clearly considerable freedom in the representation of a parallel family.

Theorem 1

If R(AB) and R(CD) are collinear and like directed, or

are parallel, then $F[R(AB)] = F[R(CD)]$.   (Ex.)

Definition

(Hyperparallel family)   The collection of lines perpendicular to line b is the <u>hyperparallel family with base line</u> b,
denoted by $F(b)$.

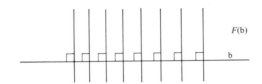

The three types of line collections, a pencil of lines
at a point, a parallel family, and a hyperparallel family are
the <u>fundamental line families</u> of hyperbolic geometry. They
are fundamental in the sense that each two lines belong to
exactly one of these three families and thus determine the
family.

To see how the three fundamental line families are
related to the 3-point problem, let us look at a circle in
absolute geometry from a new point of view. Let A and P be
distinct points in the absolute plane, with $d(A,P) = h$. Then,
by definition,

$$C(A,h) = C(A,d(A,P)) = \{X: d(A,X) = h\}. \qquad (1)$$

Now, consider any line t in the pencil $P(A)$. Because the line
reflection r_t maps the circle onto itself, it maps P to a point
Pr_t on the circle. Conversely, suppose that Q is any point
of the circle. If $Q \neq P$, then the line t that is the perpendicular

bisector of S[PQ] belongs to P(A), and $P\Gamma_t$ = Q.   If Q = P,

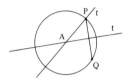

then t = L(PA) ∈ P(A) and $P\Gamma_t$ = P = Q.   Thus the circle may
be looked at as the set of all the images of P under reflections in lines of the pencil P(A), that is

$$C(A,d(A,P)) = \{P\Gamma_t: \ t \ \epsilon \ P(A)\}. \qquad (2)$$

Now consider a triangle ΔPQR in the absolute plane.
The perpendicular bisector u of S[PQ] is the unique line such
that $P\Gamma_u$ = Q, and the perpendicular bisector v of S[PR] is
the unique line such that $P\Gamma_v$ = R.   If u and v intersect at

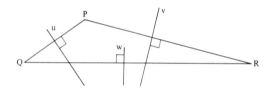

a point A then, by definition (2), Q and R belong to the circle
C(A,d(A,P)).   In that case, d(A,Q) = d(A,R) shows that the
third perpendicular bisector w of S[QR] also belongs to P(A).
Thus we have the following theorem.

Theorem 2.   (A.G.)

If two of the perpendicular bisectors of the sides of
ΔPQR belong to a pencil at point A, then so does the third,
and the vertices P, Q, R lie on a circle with center A.

The situation described above shows why the 3-point
property holds in euclidean geometry.   In that geometry, u and

v must intersect, hence the circumcircle of the triangle must exist. However, in hyperbolic geometry there is the possibility that u and v may determine any one of the three fundamental line families. If u and v do determine a pencil $P(A)$, then P, Q, and R belong to a circle with center A. On the other hand, if u and v do not determine a pencil, then P, Q, and R cannot belong to any circle.

Following the pattern suggested by the alternate definition for a circle, we now define two new figures in our geometry, and we will refer to them informally as "curves".

Definition. (Limit circle)

The set of all the images of point P under reflections in lines of the parallel family $F[R(AB)]$ is a limit circle, or horocycle, denoted by $LC[R(AB);P]$. Thus,

$$LC[R(AB);P] = \{P\Gamma_t : t \in F[R(AB)]\}.$$

The lines of the parallel family are the radial lines of the limit circle. A closed ray, with origin on the limit circle and parallel to all the radial lines that do not contain it, is a radial ray.

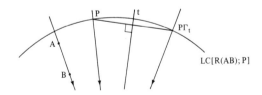

Definition. (Equidistant curve)

The set of all the images of a point P, not on line b, under reflections in the lines of the hyperparallel family $F(b)$ is an equidistant curve with base line b, denoted by $EC(b;P)$.

Thus,

$$EC(b;P) = \{P\Gamma_t : t \in F(b)\} .$$

The lines of the hyperparallel family are the <u>radial</u> <u>lines</u>
of the equidistant curve.

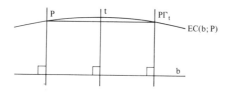

The three types of curves, circles, limit circles, and
equidistant curves form the family of hyperbolic "cycles".
What our discussion up to this point strongly suggests is that
each three noncollinear points in this geometry belong to a
cycle rather than to a circle. We will show that this suggestion is correct. However, since a careful proof is somewhat
involved, we will obtain it as part of a developmental discussion that makes the steps easier to follow.

It will expedite the discussion to first recall that a
<u>median</u> of a triangle is a segment that joins a vertex to the
midpoint of the opposite side. That the open segment of a
median is interior to the triangle, and hence to each angle of
the triangle, is a property of absolute geometry whose verification is left to the reader.

Let a general triangle be denoted by $\triangle PQR$ and let α , β
and γ be the measures of the angles at P, Q, and R respectively.
Also, let u, v, and w be the perpendicular bisectors of the
sides, with $u \perp S[PQ]$ at M, $v \perp S[PR]$ at N, and $w \perp S[QR]$ at O.

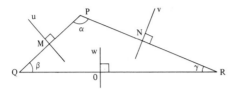

We want to show that the three lines u, v, w must belong to a
fundamental line family and hence that P, Q, R must belong to a
cycle.

One of the numbers α, β, γ, is max {α, β, γ} and we sup-
pose the notation is chosen so that α $\overset{>}{=}$ β and α $\overset{>}{=}$ γ. We first
consider the case in which α < β + γ. At P there is a ray R(PX),
in the R-side of L(PQ), such that $\not\!\!\times$QPX° = β. Because α = $\not\!\!\times$QPR°
$\overset{>}{=}$ β, R(PX) intersects S(QR] at a point A, and $\not\!\!\times$PQR $\overset{\sim}{=}$ $\not\!\!\times$PQA. The
triangle ΔPQA has congruent angles at P and Q, hence is isosceles
with base S[PQ]. Thus u = L(MA) and S[AM] is a median of ΔQPA,
so clearly S(MA) ⊂ In(ΔPQA) ⊂ In(ΔPQR). If α = γ, ΔPQR is
isosceles, with base S[PR]. In this case, v = L(NQ) and, as
T(QN) is interior to $\not\!\!\times$PQA, v must intersect S(MA). Thus u and v
intersect at a point interior to $\not\!\!\times$PQR and, by Th. 2, w must belong
to the pencil at this point. If α > γ there is a ray R(PY), in the
R-side of L(PQ), such that $\not\!\!\times$QPY = α - γ. Since α < β + γ , α - γ < β
and so R(PY) ⊂ In($\not\!\!\times$QPA) implies that R(PY) intersects S(QA) at a
point B and intersects S(MA) at a point C, and <PCB>. From R(PB)
⊂ In($\not\!\!\times$QPA) ⊂ In($\not\!\!\times$QPR), we have $\not\!\!\times$BPR° = $\not\!\!\times$QPR° - $\not\!\!\times$QPB° = α - (α - γ)
=γ, so ΔPBR is isosceles, with base S[PR], and v = L(NB). From
<QAR> and <QBA> it follows that <BAR>, and this, with <PCB>
implies $\not\!\!\times$PBR = $\not\!\!\times$CBA. Therefore R(BN) intersects S(CA) and hence
S(MA), at a point D interior to ΔPQR. By Th. 2, w also belongs to

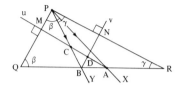

the pencil $P(D)$. Thus, we have the following result.

Theorem 3.

        If the largest angle of a triangle has measure less
than the sum of the measures of the other two angles, the triangle has a circumcircle whose center is interior to the triangle.

        With the same notations as before, we next consider the
case in which $\alpha = \beta + \gamma$. Now $\alpha > \beta$ so the ray $R(PX)$, with
$\sphericalangle QPX^{o} = \beta$, intersects $S(QR)$ at A. And since $\alpha - \gamma = \beta$, now
the ray $R(PY) = R(PX)$, and $B = A$. The triangles $\triangle QPA$ and
$\triangle RPA$ are both isosceles, with bases $S[PQ]$ and $S[PR]$ respectively.

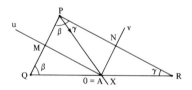

Thus $u = L(MA)$ and $v = L(NA)$ intersect at A. By Th. 2, w belongs to $P(A)$. Clearly, $A = 0$, and since A is the circumcenter
of the triangle, $S[QR]$ is a diameter, so the triangle is in-
scribed in a semicircle. Of course, $\alpha + \beta + \gamma = \alpha + \alpha = 2\alpha < 180^{o}$
implies that $\sphericalangle QPR$ is acute. We summarize these results and
also note an interesting fact whose verification is left to

the reader.

Theorem 4.

If the biggest angle of a triangle has measure equal
to the sum of the measures of the other two angles, the side
opposite to the biggest angle is a diameter of the circumcircle.

Theorem 5.

If S[QR] is a diameter of a circle, and if P on the
circle is neither Q nor R, then $\angle$QPR is acute.   (Ex)

In the triangle $\triangle$PQR with which we have been working,
the perpendicular bisectors u, v are special in the sense that
they bisect sides on the arms of a biggest angle. Our strategy
now is to make use of Th. 3 and Th. 4 to help us see what occurs if these special bisectors are parallel or hyperparallel.
Suppose, first, that u and v do not intersect. By Th. 3 and
Th. 4, u $\cap$ v = $\emptyset$ implies that $\alpha > \beta + \gamma$. Now, $\angle$QPX$^\circ$ = $\beta < \alpha$,
so A is on S(QR), and $\angle$QPY$^\circ$ = $\alpha - \gamma > \beta$, so B is on S(AR).
Let C and D be points such that <MAC> and <NBD>. The triangles

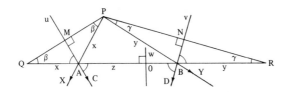

$\triangle$QPA and $\triangle$RPB are isosceles and u = L(MA) and v = L(NB).
Because u bisects $\angle$QAP, $\angle$QAM is acute, hence its opposite
angle $\angle$BAC is acute. Similarly, v bisects $\angle$RBP, so $\angle$RBN is
acute, so its opposite angle $\angle$ABD is acute.

Now suppose that u ||v. Since <MAC> and <NBD>, and
since M and N are in the P-side of L(AB), R(AM) $\not\sqcap$ R(BD), and

and R(BN) $\not\Vert$ R(AC).   Because $\not\angle$BAC and $\not\angle$ABD are acute, R(AM)
$\Vert$ R(BN) would imply that the biangle (M-AB-N) had two obtuse
angles.   Therefore R(AM) $\not\Vert$ R(BN).   Thus u $\Vert$ v implies that
R(AC) $\Vert$ R(BD).   Let x = d(A,Q) = d(A,P), y = d(B,R) = d(B,P),
and z= d(A,B).   Then d(O,Q) = d(O,R) = $\frac{1}{2}$(x + y + z).   From
the lengths of the sides in $\triangle$PAB, y + z > x, hence x + y + z
> 2x, and so d(Q,O) = $\frac{1}{2}$(x + y + z) > x = d(Q,A).   Thus O $\in$
R(AB).   Similarly, x + z > y implies that x + y + z > 2y.
Thus d(R,O) = $\frac{1}{2}$(x + y + z) > y = d(R,B), so O $\in$ R(BA), and
therefore <AOB>.   Thus the line w intersects the biangle
(C-AB-D) at O between A and B.   The line w cannot intersect
u or v because, by Th. 2, that would imply u $\cap$ v $\neq \emptyset$.   Thus
w cannot intersect either of the ray sides of (C-AB-D) and
so, by the Pasch properties of the biangle, w must be parallel
to both ray sides.   We have therefore established the following
special case, which we list for reference:

    if u $\Vert$ v, then u, v, and w belong to a parallel family. (A)

    Next, suppose that u)(v. Now there exists a line b
that is perpendicular to u at a point M_1 and to v at a point
N_1.   Line b cannot be L(PQ) since then lines L(PQ) and L(PN)
would both be perpendicular to v.   Similarly v cannot be L(PR).

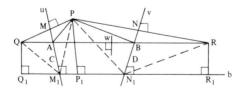

Thus b is hyperparallel to L(PQ), since both are perpendicular
to u, and b is hyperparallel to L(PR), since both are

perpendicular to v. Thus b does not intersect either S[PQ]
or S[PR]. By the Pasch property of $\triangle PQR$, it follows that
b ∩ S[QR] = $\emptyset$, and hence b ∩ S[AB] = $\emptyset$; so M_1 and N_1 lie in
one side of L(AB). If M_1 and N_1 were in the P-side of L(AB),
the acuteness of ⋨CAB and ⋨DBA would imply that the quadrilateral ABN_1M_1 had two obtuse angles as well as two right angles,
which is impossible. Therefore M_1 and N_1 are in the C-D-side
of L(AB).

Let Q_1, P_1, and R_1 denote the feet in b of P, Q, and R
respectively. Because u)(v, and <QMP> and <QAB>, u separates
P and Q, and P lies in the v-side of u. Similarly, <RNP> and
<RBA> imply that v separates P and R and that P lies in the
u-side of v. Thus $L(PP_1)$, which hyperparallel to u and v,
lies in the strip between them, and this implies $<M_1P_1N_1>$.
The lines $L(QQ_1)$, u, $L(PP_1)$, v, and $L(RR_1)$ all belong to F(b)
and the feet Q_1, M_1, P_1, N_1, and R_1 are successive on b in
that order.

The fact that M_1 is on u implies that $d(M_1,Q) = d(M_1,P)$
and that ⋨MM_1Q^o = ⋨MM_1P^o. As complements of these congruent
angles, we have⋨$QM_1Q_1^o$ = ⋨$PM_1P_1^o$, and so, by hypotenuse-angle,
it follows that $\triangle QM_1Q_1 \cong \triangle PM_1P_1$, and hence that $d(Q,Q_1) =$
$d(P,P_1)$. An entirely similar argument shows that $\triangle PN_1P_1 \cong$
$\triangle RN_1R_1$, and hence that $d(P,P_1) = d(R,R_1)$. Thus $d(Q,Q_1) =$
$d(R,R_1)$, and this implies that QQ_1R_1R is a Saccheri quadrilateral, with summit S[QR]. The altitude line is the perpendicular bisector of the summit and is therefore w. The altitude line is perpendicular to the base, so w ⊥ b, hence w is
also in F(b). Thus we have our second special result:

if u)(v then u,v,w,belong to a hyperparallel family   (B)

Now we can generalize the statements (A) and (B).

Consider either of the pairs u,w or v,w, say u,w. If u || v, then u can-
not intersect v since, by Th. 2, that would imply that u
intersected w, contradicting u || w. Similarly, u cannot be
hyperparallel to v, since, by (B), that would imply that u is
hyperparallel to w, contradicting u || w. Thus u || v and so,
by (A), u, v, and w belong to a parallel family. If u)(w
then, by the same reasoning, Th. 2 and (A) exclude u ∩ v ≠ ∅
and u || v. Therefore u)(v and so, by (B), u, v, and w belong
to a hyperparallel family. From these generalizations, and
because $P r_u$ = Q and $P r_v$ = R we have the following theorem.

Theorem 6.

        If lines u, v, w are the perpendicular bisectors of
the sides of ΔPQR and if two of these lines are parallel then
all three belong to a parallel family and are radial lines
of a limit circle that passes through P, Q, and R. If two of
the lines u, v, w are hyperparallel then all three belong to
a hyperparallel family and are radial lines of an equidistant
curve that passes through P, Q, and R.

Corollary 6. Every triangle has a circumcycle.

        The 3-point property in Cor. 6 is just one instance of
the manner in which cycles in hyperbolic geometry play the
role of circles in euclidean geometry. In particular, each
of the two new curves has a special property that is basic
to the structure of the geometry. In the remainder of this
section, we shall continue the investigation of equidistant
curves and limit circles to an extent sufficient to establish
their general nature and most characteristic properties.

        In proving the previous property (B), we already had
an indication of the next theorem which justifies the name

"equidistant curve" and states the most striking property
of such a curve.

Theorem 7.

The equidistant curve $EC(b;P)$ is the set of all
points in the P-side of b whose distance from b equals $d(P,b)$.

Proof.

Let F be the foot in b of point P and let $u = L(PF)$.
Since $u \in F(b)$, and $P\Gamma_u = P$, $P \in EC(b;P)$. If t is an arbitrary radial line in $F(b)$, $t \perp b$ implies that Γ_t maps b onto
itself. Since $X' = X\Gamma_t$ also maps the P-side of b onto itself,
$P' = P\Gamma_t$ is in the P-side of b. Also, motions preserve distance
and perpendicularity, so $d(P,F) = d(P',F') = d(P',b)$. Thus all
points of $EC(b;P)$ are at distance $d(P,F)$ from b .

Conversely, let Q be any point in the P-side of b and
such that $d(Q,b) = d(P,b) = d(P,F)$. If $Q = P$, then $Q \in EC(b;P)$.
If $Q \neq P$, then Q has a foot G in b distinct from F and,
because $d(Q,G) = d(P,F)$, QGFP is a Saccheri quadrilateral.
The line t that is the perpendicular bisector of the summit
S[QP] is the altitude line, hence is perpendicular to S[GF] ,
hence is a line in $F(b)$. Since $Q = P\Gamma_t$, $Q \in EC(b;P)$.   □

Corollary 7.1

Each radial line of the equidistant curve $EC(b;P)$ intersects the curve at exactly one point, and if $Q \in EC(b;P)$ then
$EC(b;Q) = EC(b;P)$.   (Ex.)

Definition

(Chord, secant)   The closed segment joining two points

of a cycle is a <u>chord</u> of the cycle. The line of a chord is a
<u>secant</u> of the cycle.

Corollary 7.2

The perpendicular bisector of a chord of an equidistant
curve is a radial line of the curve, and the secant of the chord
is hyperparallel to the base line.   (Ex.)

Corollary 7.3

No line intersects an equidistant curve at more than two
points.   (Ex.)

If $d(P,b) = h > 0$, the reflection in line b, namely $X' = X\Gamma_b$ maps the curve $EC(b;P)$ to a congruent equidistant curve
$EC(b;P')$ in the opposite side of b. Some texts define an equi-
distant curve, of distance h, to be the set of all points at
distance h from a line b. Under that definition, the curve is
$EC(b;P) \cup EC(b;P')$ and hence, like the euclidean hyperbola, is
a curve with two "branches". However, we will refer to
$EC(b;P) \cup EC(b;P')$ as "the union of the two equidistant curves
at distance h from b".

That an equidistant curve is "concave" toward its base
is expressed by the following property (implied by Th. 6, III-4).

Theorem 8

If P and Q are two points of an equidistant curve with
baseline b and if u and v are the radial lines through P and Q
respectively, all points of the curve between u and v lie in
the non-b-side of $L(PQ)$.   (Ex.)

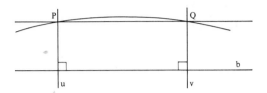

Turning next to a limit circle LC[R(AB);P], we first
observe that, by Th. 1, there exists a ray R(PC) such that the
parallel family F[R(AB)] is also represented by F[R(PC)].
Thus the limit circle LC[R(AB);P] is also the limit circle
LC[R(PC);P], and we can obtain a simpler way of representing
the curve by the following convention.

Convention

The notation "LC[R(PC)]" will be understood to have the
meaning of "LC[R(PC);P]".

To show that a limit circle LC[R(PA)] has properties
similar to the equidistant curve properties in Cor. 7.1, 7.2
and 7.3, we will use a 'stepwise' argument. First, because
L(PA) is in the family F[R(PA)], and since a reflection in L(PA)
maps P onto itself, P $\in$ LC[R(PA)]. Next, if Q on the curve is
not P, then by the definition of Q $\in$ LC[R(PA)] there exists a
radial line u such that Pr_u = Q, so u must be the perpendicular
bisector of S[PQ]. Therefore Q is not on L(PA) since that would
imply that u was perpendicular to L(PA) and parallel to R(PA).
Thus we have shown that:

$$L(PA) \cap LC[R(PA)] = \{P\} . \qquad\qquad (i)$$

Next, suppose that the secant L(PQ) intersects the limit circle
at a third point R. Then there exists a radial line v such

P Γ_v = R, so v is the perpendicular bisector of S[PR].   But
since u and v are perpendicular to L(PQ), they are hyperparallel
and this contradicts the property that radial lines are parallel.
The contradiction implies that:

no line through P intersects LC[R(PA)] more than twice. (ii)

We now want to show that if Q on LC[R(PA)] is not P, and
if R[QB] is the radial ray at Q, then LC[R(PA)] = LC[R(QB)] .
We know, from Th. 1, that F[R(PA)] and F[R(OB)] represent the
same parallel family.   What we must show is that under reflections in the lines of this family, all the images of P form the
same set as do all the images of Q.   If u is the radial line
such that PΓ_u = Q then u is also a radial line such that
Q Γ_u = P, so P ∈ LC[R(QB)].   Also, from (i), Q ∈ LC[R(QB)].
Let R denote any point of LC[R(PA)] which is neither P nor Q,
and let v be the radial line such that PΓ_v = R.   By property
(ii), R is not on L(PQ), so ΔPQR exists.   Because u is the

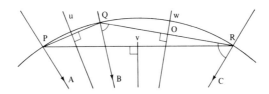

perpendicular bisector of side S[PQ] and v is the perpendicular
bisector of side S[PR], and u ||v, it follows from Th. 6 that
w, the perpendicular bisector of S[QR], belongs to F[R(QB)].
And since   QΓ_w = R, it follows that R ∈ LC[R(QB)].   Thus we
have shown that LC[R(PA)]⊂LC[R(QB)].   But since Q and P belong
to LC[R(PA)] ∩ LC[R(QB)], the same argument, with R regarded as
a third point of LC[R(QB)], shows that LC[R(OB)] ⊊ LC[R(PA)].

The two inclusions imply that

$$LC [R(PA)] = LC [R(QB)].   \text{(iii)}$$

In the argument for (iii), Q was taken to be any point
of LC[R(PA)] distinct from P. Thus (iii) and (i) imply that
no radial line intersects the curve more than once, and (iii)
and (ii) imply that no secant intersects the curve more than
twice. Also, it was shown that w, the perpendicular bisector
of the chord S[QR], is a radial line of the limit circle. If
O is the midpoint of S[QR], and $C = B \Gamma_w$, the reflection in w
maps ≮OQB onto ≮ORC. Since $w \perp L(QR)$ and $w \, || \, R(PA) \, || \, R(OB)$,
it follows that $≮OQB^{\circ} = \pi[d(O,Q)]$. Therefore $d(O,Q) = d(O,R)$
and $≮OQB^{\circ} = ≮ORC^{\circ}$ imply that $≮ORC^{\circ} = \pi[d(O,R)]$ and hence that
$R(RC) \, || \, w$. Because Γ_w maps the B-side of L(QR) onto itself,
$R(RC) \, || \, w$ implies that $R(RC) \, || \, R(QB) \, || \, R(PA)$. Thus R[RC)]is
a radial ray, and (B-QR-C) is an isosceles biangle.

The next theorem summarizes results from the preceding
discussion.

Theorem 9

No radial line of a limit circle intersects the curve
more than once and no secant intersects the curve more than
twice. If R [QB) is a radial ray of the limit circle LC[R(PA)],
then LC[R(QB)] = LC [R(PA)]. The perpendicular bisector of a
chord of a limit circle is a radial line and the union of a
chord with the radial rays at its endpoints is an isosceles
biangle.

We have not yet proved that every radial line of a
limit circle LC[R(PA)] intersects the limit circle. That is,
we have not shown that every line in the family F[R(PA)] is
the image of L(PA) under reflection in some line of the family.

We do so in our next theorem.

Theorem 10

If t || R(PA), then t intersects LC[R(PA)] at exactly
one point.

Proof

By definition, t ||R(PA) implies the existence of a ray
R(QB) ⊂ t and such that R(QB) ||R(PA). This last parallelism
implies that the ray R(PX) that bisects ⦨APQ intersects R(QB)
at a point C. The ray that bisects ⦨PQB (=⦨PQC) intersects
S(PC) at a point D, and D ∈ In(A-PQ-B). Let F, G, and H be the
feet of D in the lines L(PA), L(QB) and L(PQ) respectively.
Since ⦨APD° = ½⦨APQ° < 90°, F ∈ R(PA), and we may suppose
<PFA>. Similarly, ⦨BQD° < 90° implies that G ∈ R(QB) and we may
suppose <QGB>. The point H belongs to R(PQ) ∩ R(QP), hence is

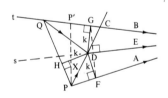

between P and Q. Because R(PD) bisects ⦨APQ, d(D,F) = d(D,H),
and because R(QD) bisects ⦨BQP, d(D,H) = d(D,G). Thus d(D,F) =
d(D,H) = d(D,G) = k.

Since D ∈ In(A-PQ-B), there exists a ray R(DE) parallel
to both R(PA) and R(QB). From <PFA> and <QGB>, it follows that
R(DE) is parallel to both R(FA) and R(GB). The right biangles

(E-DF-A) and (E-DG-B) are congruent, and $\sphericalangle FDE^{\circ} = \sphericalangle GDE^{\circ} = \pi(k)$ < 90°. Thus $\triangle FDG$ exists and is isosceles with base S[FG]. The line L(DE) = s is the perpendicular bisector of S[FG], so the line reflection Γ_s maps F to G and maps the right angle $\sphericalangle DFA$ onto the right angle $\sphericalangle DGB$. Thus Γ_s maps L(PA) onto t. Because s || R(PA), s is a radial line of LC[R(PA)], hence P' = $P\Gamma_s$ belongs to t ∩ LC[R(PA)] and, from Th. 8, {P'} = t ∩ LC[R(PA)].   □

Corollary 10

If R(QB) || R(PA), the point Q is a point of LC[R(PA)] if and only if (A-PQ-B) is isosceles.   (Ex.)

We now want to prove that all limit circles are congruent. A reason for suspecting this property also suggests how it might be proved. If R[AB] and R[CD] are any two closed rays, then, as one would suppose, there exists a motion which maps R[AB] onto R[CD]. Since R[AB] determines LC[R[AB]] and R[CD] determines LC[R[CD]], a natural guess is that the motion also establishes a congruence of these limit circles. But to verify this, we need another property of motions.

Definition

(Product of motions) If Γ_1 and Γ_2 are motions of the absolute plane, the product of Γ_1 and Γ_2 in that order is the mapping Γ, also denoted by $\Gamma_1\Gamma_2$, and defined by
$$X\Gamma = (X\Gamma_1)\Gamma_2.$$

Theorem 11

(A.G.) The product of two motions of the absolute

plane is itself a motion of the absolute plane.

Proof

Let Γ_1 and Γ_2 denote motions of the plane A^2 and let
$\Gamma = \Gamma_1\Gamma_2$. If $X' = X\Gamma$, and $X^* = X\Gamma_1$, then, by definition,
$X' = (X\Gamma_1)\Gamma_2 = X^*\Gamma_2$. Because Γ_1 is a motion, $d(X,Y) =$
$d(X^*,Y^*)$, and because Γ_2 is a motion, $d(X^*,Y^*) = d(X',Y')$.
Thus $d(X,Y) = d(X',Y') = d(X\Gamma,Y\Gamma)$ shows that Γ is an isometry.
Because the correspondence $X \to X^*$ is one-to-one, and the correspondence $X^* \to X'$ is one-to-one, the correspondence $X \to X'$
is also one-to-one. Thus Γ is a one-to-one, distance preserving mapping of A^2 onto itself and is therefore a motion of A^2.

Theorem 12

(A.G.) Two closed rays are congruent.

Proof

Let $R[AB)$ and $R[CD)$ be distinct rays. We may suppose
the representation $R[CD)$ is chosen so that $d(A,B) = d(C,D)$. By
Th. 3, II-6, the image of $R[AB)$ in any motion is again a closed
ray. In particular, the image ray will be $R[CD)$, and the
theorem will be proved, if the motion maps A onto C and B onto
D.

Case 1

The rays have the same origin, i.e., $A = C$. Then $B \neq D$,
since $R[AB) \neq R[CD)$, and line s, the perpendicular bisector of
$S[BD]$, exists. Because $d(A,B) = d(C,D) = d(A,D)$, A is

equidistant from B and D and is therefore on s. Thus the
line reflection Γ_s maps B onto D and A onto A = C.

Case 2

The rays have different origins, i.e. A ≠ C. Now
line u, the perpendicular bisector of S[AC] exists, and the
line reflection X* = $X\Gamma_u$ maps A onto C and maps B to B*;
therefore d(A,B) = d(A*,B*) = d(C,B*). If B* = D, then Γ_u
is a motion that maps A to C and B to D. If B* ≠ D, then a
line v that is the perpendicular bisector of S[B*D] exists.
Because d(C,B*) = d(A,B) = d(C,D), C is equidistant from B*
and D, and so C is on v. Thus $C\Gamma_v$ = C and $B*\Gamma_v$ = D. But
now we have $(A\Gamma_u)\Gamma_v = C\Gamma_v$ = C, and $(B\Gamma_u)\Gamma_v = B*\Gamma_v$ = D. Thus
$\Gamma = \Gamma_u\Gamma_v$ is a motion that maps A onto C and B onto D.   □

Corollary 12

If S[AB] ≅ S[CD], there exists a motion of the plane Γ
such that AΓ = C and BΓ = D.

Theorem 13

A motion of the hyperbolic plane maps a biangle onto
a biangle.   (Ex.)

Theorem 14

A motion Γ of the hyperbolic plane that maps R[AB)
onto R[CD) also maps LC[R(AB)] onto LC[R(CD)].

Proof

Without loss of generality, we may suppose that $d(A,B)$
= $d(C,D)$, hence that $X' = X\Gamma$ maps A to $A' = C$ and maps B to
$B' = D$. If P is any point of $LC[R(AB)]$, distinct from A,
and $R[PP_1)$ is the radial ray at P, then, by Th. 9, $(B-AP-P_1)$
is an isosceles biangle. Since Γ maps $(B-AP-P_1)$ to a congruent biangle, (Th. 13), $(B'-A'P'-P_1')$ is isosceles, and this,
by Cor. 10, implies that $P' \in LC[R(CD)]$. Thus the Γ-image
of $LC[R(AB)]$ is contained in $LC[R(CD)]$.

Now, let Q be any point of $LC[R(CD)]$ and let $R[QQ_1)$
be the radial ray at Q. If $Q = C$, then Q is the Γ-image of
A. If $Q \neq C$, then by Th. 9, the biangle $(D-CQ-Q_1)$ is isosceles. The inverse of Γ, the mapping $X* = X\Gamma^{-1}$, is a motion
that maps C to $C* = A$ and maps D to $D* = B$ and maps $(D-CQ-Q_1)$
to an isosceles biangle $(B-AQ*-Q_1^*)$. Because $(B-AQ*-Q_1^*)$ is
isosceles, it follows from Cor. 10 that $Q* \in LC[R(AB)]$. But,
by the definitions of Γ and Γ^{-1}, $Q* = Q\Gamma^{-1}$ implies that $Q =$
$Q*\Gamma$. Thus every point of $LC[R(CD)]$ is the Γ-image of a point
in $LC[R(AB)]$, and therefore Γ maps $LC[R(AB)]$ onto $LC[R(CD)]$.

Corollary 14

The limit circles $LC[R(AB)]$ and $LC[R(CD)]$ are congruent.

Proof

By Th. 12, there is a motion that maps $R[AB)$ onto
$R[CD)$ and, by Th. 14, this motion also maps $LC[R(AB)]$ onto
$LC[R(CD)]$. Therefore $LC[R(AB)] \cong LC[R(CD)]$.   □

That a limit circle is concave toward the directions
of its radial rays can be expressed the following way.

Theorem 15

If $R[AA_1)$ and $R[BB_1)$ are radial rays of a limit circle, then all points of the limit circle between the radial
lines $L(AA_1)$ and $L(BB_1)$ lie in the non-A_1-B_1-side of the
line $L(AB)$.

Proof

Let $R[PP_1)$ be a radial ray such that $L(PP_1)$ is between
$L(AA_1)$ and $L(BB_1)$. By Cor. 12, III-1, $L(PP_1)$ separates A and
B and therefore intersects $S(AB)$ at some point N, and, by Th.
9, $N \neq P$. Assume that $P \in A_1$-B_1-side of $L(AB)$. Then $R(PP_1)$ ||
$R(AA_1)$ || $R(BB_1)$ implies that $<NPP_1>$ and $P \in In(A_1$-AB-$B_1)$.

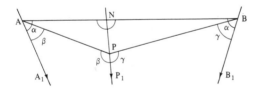

Let α, β, and γ be the measure of the angles in the isosceles
biangles $(A_1$-AB-$B_1)$, $(P_1$-PA-$A_1)$ and $(P_1$-PB-$B_1)$ respectively.
Since $P \in In(A_1$-AB-$B_1)$, $\alpha > \beta$ and $\alpha > \gamma$. Because $<NPP_1>$,
$\angle P_1PA$ is exterior to $\triangle PNA$, and therefore $\beta > \angle PNA^{\circ}$. Similarly,
$\angle P_1PB$ is exterior to $\angle PNB$ and therefore $\gamma > \angle PNB^{\circ}$. Thus,
$\beta + \gamma > \angle PNA^{\circ} + \angle PNB^{\circ} = 180^{\circ}$ and this imples that $2\alpha > \beta + \gamma >$
180°. But this is impossible since the angle sum of $(A_1$-AB-$B_1)$,
which is 2α, is less than 180°. Thus P cannot be in the
A_1-B_1-side of $L(AB)$, and since $P \neq N$, P must be in the non-A_1-
B_1-side of $L(AB)$. □

By now it should be clear that in hyperbolic geometry

there is a pattern of properties common to circles, limit circles, and equidistant curves. Corollary 6 was an example of
a cycle theorem. The following are three more examples.

Theorem 16

No line intersects a cycle at more than two points.

(Ex.)

Theorem 17

The perpendicular bisector of a chord of a cycle is
a radial line.   (Ex.)

Theorem 18

Each radial line of a cycle is an axis of symmetry
to the cycle.   (Ex.)

The definition of a tangent line to a general curve
in H^2 involves the use of limits. For the special case of
cycles, however, we can use special properties that distin-
quish the lines it is natural to regard as the tangents to
the cycle.

Definition

(Tangent line to a cycle) A line t that intersects
a cycle at exactly one point and which does not separate any
two points of the cycle is a tangent line to the cycle.

If P is the intersection point, t is <u>tangent to the cycle at</u> P.

To discover which lines are tangent to a cycle c ,
consider a point P on c , and let s denote the radial line through
P and let t be the line perpendicular to s at P. Because t ⊥ s,
Γ_s maps t onto t and, by Th. 18, maps c onto c. Thus if t
intersected c at a second point Q, then P, Q, and Q Γ_s would be
three points of c on t, contradicting Th. 16. Therefore t
intersects c only at P.

Now, let Q_0 denote a point of c that is not on s and let
R(PA) be the open ray at P on s that is in the Q_0-side of t.

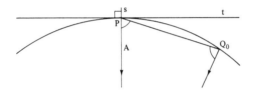

For all three types of cycles, the angle ∢APQ_0 is acute since it
is a base angle of an isosceles triangle (for c a circle), an
angle of an isosceles biangle (for c a limit circle), or a summit
angle of an S-quadrilateral (for c an equidistant curve). As
a point Q, not on s, varies on c, the acuteness of ∢APQ implies
that Q is in the Q_0-side of t. In the case of a circle, the
second point of c on s is clearly also in the Q_0-side of t.
Thus the closed Q_0-side of t contains c, and so t does not separate any two points of c. Therefore t is a tangent to c at P.

Because there are points of c not on s, and $c\Gamma_s = c$,
s does separate some points of c and is therefore not a tangent.
To see that t is the unique tangent at P, let u be any line of

the pencil at P that is distinct from s and t. Then there

exists a ray R(PW) ⊂ u and such that ∡APW^o = α < 90°.

If c is a circle C(B,r), B has a foot F in u that

belongs to R(PW), and d(B,F) < r implies that F is interior

to the circle. Therefore, by Th. 1, II-5, u intersects the

circle at a second point and is a secant, not a tangent.

If c is the limit circle LC[R(PA)], let M be the point

on R(PW) such that d(P,M) = $\pi^{-1}(\alpha)$. By Cor. 3.1, III-4, the

line v that is perpendicular to u at M is parallel to R(PA)

and is therefore a radial line of LC[R(PA)]. By the definition

of the limit circle, PΓ_v is on c and is also on u, since uΓ_v =

u. Thus u is a secant, not a tangent.

Finally, if c is an equidistant curve EC(b;P), s intersects

b at a point F. Let R(PC) denote a ray parallel to b and let

R(PD) be the open ray on t in the C-side of s. The reflection

X' = XΓ_s maps C to C' and D to D'; ∡CPC' is the fan angle of

P and b, and R(PD') is the open ray opposite to R(PD). If u

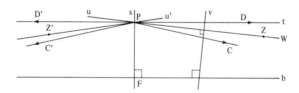

subdivides ∡CPD, then u)(b and there exists v perpendicular to

both b and u. Thus v is a radial line of c, and since ∡FPW is

acute, v is in the C-D-side of s. By the definition of the

cycle, the reflection in v maps P to a point Z on c, so u is a

secant. The mapping Γ_s maps u to u' subdividing ∡C'PD' and

maps Z to a point Z' on $c \cap u'$, hence u' is also a secant. The
lines L(PC) and L(PC') and all lines subdividing $\angle$CPC' clearly
separate Z and Z' and hence are not tangents. In all cases,
then, u is not a tangent.

From the arguments just given, we have the following
characterization of the tangents to a cycle.

Theorem 19

The line that intersects a cycle at a point P and is
perpendicular to the radial line through P is the unique line
tangent to the cycle at P.

We have not, as yet, given any explanation for the name
"limit circle". Let(A_1-AB-B_1) be an isosceles biangle and let

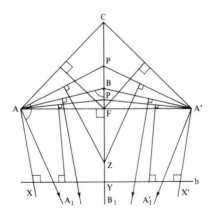

F be the foot of A in $L(BB_1)$ and C on R(FB) be such that d(A,F)
equals d(F,C). The reflection in $L(BB_1)$ maps R(AA$_1$) to R(A'A$_1'$),
and A, B, and A' lie on a limit circle. Now consider P variable
on S[CB]. When P = C, the perpendicular bisectors of S[AC]
and S[A'C] intersect at F, the circumcenter of $\triangle$ACA'. As P
moves towards B, the perpendicular bisectors of S[AP] and S[A'P]
intersect at a point Z on $R(BB_1)$ that moves farther and farther

from B. The triangles ΔAPA' have circumcircles with center Z.
As d(B,P) decreases, the radius of the circle through A, P, A'
increases, and the circle 'approaches' LC[R(AA₁)], hence the
name "limit circle".

Now, consider P variable on S(FB). The angle ∢FPA is
acute and there exists a ray R(AX) in the F-side of L(PA) and
such that ∢PAX° = ∢APF°. The lines L(PF) and L(AX) are hyperparallel and have a common perpendicular b, intersecting L(PF)
at Y in the F-side of L(AP). The three points A, P, A' lie on
an equidistant curve EC(b;P). As P moves toward B, the angle
∢YPA decreases and the distance d(P,Y) increases. The base
line b moves farther and farther away from F and the equidistant curves 'approach' the limit circle. As P moves in the
opposite direction toward F, the ∢FPA approaches 90°, b moves
toward L(AA'), and the equidistant curves 'approach' the line
L(AA').

The intuitive suggestion of this admittedly informal
description is that all circles are more curved than a limit
circle and all equidistant curves are flatter than a limit
circle.

We conclude this section with an extension to cycles of
some basic concepts that are familiar in connection with circles.
The notion for example of circles with the same center, that is
concentric circles, has the following more general form for
cycles.

Definition

(Co-radial cycles)   Two cycles are co-radial if they
have the same family of radial lines.

We note, in particular, that two circles have the same
center if and only if they have the same family of radial
lines, hence for circles the terms "concentric" and " co-
radial" are synonymous.

The notion of regions interior and exterior to a circle
can also be extended to cycles. The region interior to a
cycle can be defined - for all three types - as the intersection set of all the open half-planes which intersect the cycle
and have an edge that is tangent to the cycle. However, we
already have a special definition for the sets interior and
exterior to a circle and we will adopt special definitions for
the limit circle and equidistant curve.

Definition

(Interior and exterior to a limit circle) A point is
interior to a limit circle LC[R(AB)] if it belongs to a radial
ray of the limit circle but does not belong to the limit circle. A point is exterior to the limit circle if it is neither
interior to the limit circle nor on the limit circle. The set
of all points interior to LC[R(AB)] will be denoted by
In[LC[R(AB)]] and the exterior to LC[R(AB)] will be denoted
by Ex[LC[R(AB)]].

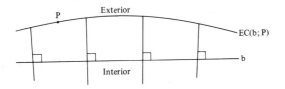

Definition

(Interior and exterior to an equidistant curve)   A
point is _exterior to_ an equidistant curve EC(b;P) if it lies
in the P-side of b and is further from b than P.   A point
is _interior to_ the equidistant curve if it is neither exterior
to the curve nor on the curve.   The sets of points interior
and exterior to EC(b;P) will be denoted respectively by
In[EC(b;P)]and Ex[EC(B;P)].

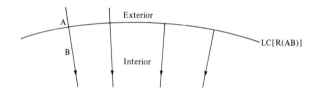

Exercises - Section 5

1.   Prove Th. 1.

2.   If line t belongs to the hyperparallel family $F(b)$,
      and if s)(t, must s belong to $F(b)$? Explain.

3.   In ΔABC, M is the midpoint of S[BC], so S[AM] is a
      median of the triangle. Prove that X ∈ S(AM) implies
      that X ∈ In(ΔABC), (see Th. 6, II-2).

4.   Prove Th. 5.

5.   Prove Cor. 7.1.

6.   Prove Cor. 7.2.

7.   Prove Cor. 7.3.

8.   Prove Th. 8.

9.   Prove Cor. 10.

10.   Prove Th. 13.

11.   Prove Th. 16.

12.   Prove Th. 17.

13.   Prove Th. 18.

14.   A circle whose center is interior to $\triangle$ ABC and which is tangent to L(AB), L(BC), and L(CA) is an <u>incircle</u> to the triangle. Explaing why in absolute geometry every triangle has a unique incircle. Does every biangle have a unique incircle? Explain.

15.   In Ex. 14, Sec. 3, it was stated that if a quadrilateral ABCD is inscribed in a circle, then $\star A^\circ + \star C^\circ = \star B^\circ + \star D^\circ$. Show that this property also holds if ABCD is inscribed in either a limit circle or an equidistant curve.

16.   If $C(A,r_1)$ and $C(A,r_2)$ are co-radial circles, with $r_1 < r_2$, and if R[AX) intersects the circles at B_1 and B_2 respectively, and R[AY) intersects them at C_1 and C_2 respectively, then $d(B_1,B_2) = d(C_1,C_2) = r_2-r_1$. Thus the circles "intercept" congruent segments on radial rays. Show that two co-radial limit circles also intercept congruent segments on radial rays. Is there a similar property for two co-radial equidistant curves? Explain.

17.   Given a point P on line t and a point Q, not on t, show that there exists a cycle which is tangent to t at P and which passes through O. Under what conditions is the cycle a circle? A limit circle? An equidistant curve?

18.   Corresponding to $\triangle$ABC, let t_1, t_2, t_3 be the altitude

lines through A,B, and C respectively. Let lines s_1, s_2, s_3 be such that $s_1 \perp t_1$ at A, $s_2 \perp t_2$ at B, and $s_3 \perp t_3$ at C. Show that if two of the lines s_1, s_2, s_3 intersect then the three altitude lines must belong to one of the three fundamental line families.

19.   Two <u>cycles</u> <u>are</u> <u>tangent</u> at a point if they intersect at that point and have a common tangent line at the point. If a circle $C(A,r)$ is contained in the interior of a right angle, $\angle BCD$, how many cycles which are tangen to $L(BC)$ at C are also tangent to $C(A,r)$? What determines whether such a cycle is a circle, a limit circle, or an equidistant curve?

Section 6. Hyperbolic Compass and Straight Edge Constructions

In the early development of euclidean plane geometry,
nearly all the figures considered consisted of segments and
circular arcs. Thus in drawing diagrams to represent these
figures, the natural instruments were a straight edge and
compass. From this association of the conceptual mathematical
figures and the physical drawings there arose a theory of
compass and straight edge constructions. In this section, we
want to indicate how a counterpart to that theory can be developed in hyperbolic geometry. Our interest in doing so does
not stem from the intrinsic importance of constructions. In
either euclidean or hyperbolic geometry, compass and straight
edge constructions form a special and minor topic in the general theory. However, the method of obtaining hyperbolic constructions is instructive, and the particular constructions
that we will consider involve geometric relations of general
importance.

Since mathematical geometry has no axioms about diagrams
or drawing instruments, what is called a "straight edge and
compass construction" is a theorem about the existence of certain points, lines, and circles which satisfy conditions that
define the theorem to be a "construction". These defining
"rules" can be stated as follows. If A and B are two given
points, the line $L(AB)$ is defined to be "constructible" or
"known", and the circles $C(A, d(A,B))$ and $C(B, d(A,B))$ are defined
to be constructible. If two known lines, or two known circles,
or a known line and known circle, intersect, then the points of
intersection are defined to be constructible. Starting with a

given set of points G, each two points A,B in G determine a
constructible line L(AB) and constructible circles C(A,d(A,B))
and C(B,d(A,B)). The intersection points of all such lines
and circles are defined to be "constructible from G" and the
set of all these intersection points form a set G_1 that contains G. In turn, each two points of G_1 determine a constructible
line and two constructible circles and the intersections of such
lines and circles is a set G_2, which is also constructible from G and
which contains G_1. Thus corresponding to the given set G,
there is a sequence of sets G_1, G_2,... whose points are constructible from G. A figure S is defined to be constructible
from G if one of the sets G, G_1, G_2,...contains a subset of
S that uniquely determines S.* To prove that S is constructible from G is to prove that one of the sets G_i contains such a
subset of S. A description of the successive steps by which
this subset can be obtained is called "a construction for S
from G".

We now want to prove a few basic construction theorems
in absolute geometry. The discussion which provides a proof
will also indicate a construction for what we show to be constructible. We begin with two given points A,B. A midpoint
M to S[AB] exists and a line t perpendicular to L(AB) at M
exists. Because d(A,B) is greater than d(A,M) = d(A,t), then,
by Cor. 1, II-5, the circle C(A,d(A,B,)) intersects t at two
points P,Q and M is the midpoint of S[PQ]. Because P is on

---

*
In the classical theory of compass and straight edge constructions, G is restricted to a pair of points at unit distance.

t, $d(P,B) = d(P,A) = d(A,B)$. Similarly, $d(Q,B) = d(Q,A) = d(A,B)$. Thus P and Q are on both the circles $C(A,d(A,B))$ and $C(B,d(A,B))$. Because these circles are constructible from A and B, the intersection points P,Q are constructible. Therefore $t = L(PQ)$ is constructible and so $M = t \cap L(AB)$ is constructible. We list this fact for reference.

Theorem 1

(A.G.) Given two points A,B, the midpoint of $S[AB]$ is constructible and the perpendicular bisector of $S[AB]$ is constructible.

Again, if A and B are two given points, the known line $L(AB)$ intersects the known circle $C(A,d(A,B))$ at a second point C. Since A is the midpoint of $S[BC]$, $C = B\Gamma_A$.   Thus we have the following theorem.

Theorem 2

(A.G.) Given two points A, B, the image of B in the reflection Γ_A is constructible.

We can apply Th. 2 to obtain a companion theorem.

Theorem 3

(A.G.) Given three noncollinear points P, A, and B, the foot of P in $L(AB)$ is constructible and the image of P under reflection in $L(AB)$ is constructible.

Proof

A point in F that is the foot of P in $L(AB)$ exists. By Th.2

the point $P' = P\Gamma_A$ is constructible, and $<PAP'>$ implies that
$d(P,P') > d(P,A) \gtreqless d(P,F) = d(P,s)$, where $s = L(AB)$. By Cor. 1,
II-5, the known circle of $C(P,d(P,P'))$ intersects s at two points
C,D, and the midpoint of S[CD] is F. Because C and D are
constructible, if follows from Th. 1 that F is constructible,
and therefore L(PF) is constructible. By Th. 2, $P\Gamma_F$ is constructible. But since L(PF) $\perp$ s at F. $P\Gamma_F = P\Gamma_s$, and so $P\Gamma_s$
is constructible.   □

Theorem 4

        (A.G.) Given two points A,B, the line perpendicular
to L(AB) at A is constructible. (Ex.)

Theorem 5

        (A.G.) Given three non-collinear points A,B,C, the
angle bisectors of $\triangle ABC$ are constructible, as are the medians
and the perpendicular bisectors of the sides.   (Ex.)

        A positive distance h is given if two points A,B are
given such that $d(A,B) = h$. Using either point or line reflections we can "copy" distances in the following sense.

Theorem 6

        (A.G.) Given two points A,B and given point P on a given
line t, the circle $C(P,d(A,B))$ is constructible as are the two
points on t at distance $d(A,B)$ from P.

Proof

        If P is either A or B, the theorem is trivially cor-
rect, so we suppose that P is neither A or B. The midpoint

M of S[PA] is constructible, (Th. 1), and the point B' =
BΓ_M is constructible. Since P = AΓ_M = A', d(P,B') =
d(A',B') = d(A,B). Therefore the constructible circle
C(P,d(P,B') is the circle C(P,d(A,B)) and this circle intersects
t at two points whose distance from P is d(A,B).   □
       The familiar euclidean construction for "copying an
angle" actually belongs to absolute geometry. Given an angle
∢ABC and a ray R[PQ), the circle C(B,d(B,A)) is constructible

and intersects R(BC) at a constructible point D. By Th. 6,
the circle C(P,d(B,A)) is constructible and intersects R(PQ)
at a constructible point E. The circle C(E,d(A,D)) is constructible. By Cor. 12, III-5, there is a motion Γ that
maps B onto P and A onto E. This motion must also map C(B,d(B,A))
onto C(P,d(B,A)) and must map C(A,d(A,D)) onto C(E, d(A,D)).
Because the first two circles intersect at D, their
Γ-image must also intersect at a point R that is therefore
constructible. Because ΔBAD $\cong$ ΔPER, by side-side-side, ∢EPR
$\cong$ ∢ABD. Thus we have the following theorem.

Theorem 7

       (A.G.) Given an angle ∢ABC and a ray R[PQ), an angle
∢QPR is constructible that is congruent to ∢ABC.
       Turning now to hyperbolic geometry, there are three

constructions which are obviously of paramount interest:
(i) given a line t and a point P not on t, construct the fan
angle of P and t; (ii) given an acute angle ∢ABC of measure
α, construct a segment of length $\pi^{-1}(\alpha)$; (iii) given two
hyperparallel lines s and t, construct the point F on s and
the point G on t such that L(FG) is the common perpendicular
to s and t. We will show that a solution to (iii) can be
obtained from a rather remarkable theorem due to J.T. Hjelmslev.
We will also make use of a corollary to the Hjelmslev theorem
in solving (i). Finally, we will apply (i) and (iii) to obtain
a solution to (ii).

　　　　As background for the Hjelmslev theorem, we need to distinguish two different senses in which a motion of the absolute
plane may map a line s onto a non-intersecting line t. If a
motion Γ maps Y and Z on s onto Y' and Z' respectively on t,
then of course Γ maps R[YZ) onto R[Y'Z'). But R(YZ) and R(Y'Z')
may either lie in one side of L(YY') or they may be separated
by L(YY'). It seems fairly obvious that whichever of these two

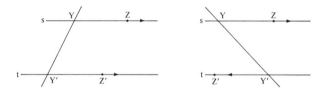

situations occurs for one pair Y,Z on s, the same situation
will occur for all pairs Y,Z on s, but we need to verify that
this is so. We first introduce some convenient names.

Definition

　　　　(Equioriented congruence) If s and t are non-intersecting

lines and if $X' = X\Gamma$ is a motion of A^2 mapping s onto t,
then an ordered pair of distinct points (A,B) on s and the
corresponding ordered pair (A',B') on t are equioriented if
B and B' are in one side of L(AA'), or equivalently if R(AB)
and R(A'B') lie on one side of L(AA'). If (Y,Z) and (Y',Z')
are equioriented for all Y,Z on s, Y ≠ Z, then the congruence
of s with t, under Γ, is an equioriented congruence.

In the next three theorems, we suppose that s and t
are two non-intersecting lines and that $X' = X\Gamma$ is a motion
of the plane A^2 mapping s onto t.

Theorem 8

(A.G.) If (A,B) and (A',B') are equioriented, then
(B,A) and (B',A') are equioriented.

Proof

Because B and B' are in one side of L(AA'), it follows
that L(AA') ∩ S[BB'] = ∅ and therefore that S[AA'] ∩ S[BB'] =
∅. Since A ≠ B and A' ≠ B', A and A' are either in one side
of L(BB') or they are in opposite sides. If they are in opposite sides, then L(BB') intersects S(AA') at a point P, and,
by Th. 5, II-4, <APA'> implies that <BPB'>. Thus
P ∈ S(AA') ∩ S(BB'), which contradicts S[AA'] ∩ S[BB']= ∅.
Thus A and A' must be in the same side of L(BB').   □

Theorem 9

(A.G.) If A, B, C are distinct points on s, the equiorientation of (A,B) and (A',B') and that of (B,C) and (B',C')

implies that (A,C) and (A',C') are also equioriented.

Proof

The equiorientation of (A,B) and (A',B') implies that
there is a B-B'-side of L(AA'). If C ∈ R(AB), then C' ∈ R(A'B')
because R(AB) Γ = R(A'B'). Since R(AB) and R(A'B') are in
the B-B'-side of L(AA'), it follows that C and C' are in this
side of L(AA'). If C ∉ R(AB), then <BAC>. Because Γ preserves
betweeness, <BAC> implies <B'A'C'>, and so C and C' are both in
the non-B-B'-side of L(AA').   □

Theorem 10

(A.G.)   If there exist distinct points A, B on s such
that (A,B) and (A',B') are equioriented, then the congruence of
s with t, under Γ , is an equioriented congruence.

Proof

The set of all ordered pairs of distinct points on s is
the union of the three sets { (A,Z): Z ≠ A } , { (Y,A): Y ≠ A },
and {(Y,Z): Y ≠ A, Z ≠ A, Y ≠ Z } .

Consider a pair (A,Z), Z ≠ A. Because (A,B) and (A',B')
are equioriented, there is a B-B'-side of L(AA'). By exactly
the same argument as in the proof of Th. 9, Z ∈ R(AB) implies that
Z and Z' are in one side of L(AA') and Z ∉ R(AB) implies that
Z and Z' are in the other side of L(AA'). Therefore (A,Z) and
(A',Z') are equioriented.

Next, if (Y,A) is an ordered pair such that Y ≠ A, then
(A,Y) is an ordered pair such that Y ≠ A. By the argument of
the preceding paragraph, (A,Y) and (A',Y') are equioriented,

hence by Th. 8, (Y,A) and (Y',A') are equioriented.

Finally, consider a pair (Y,Z), Y ≠ A, Z ≠ A, Y ≠ Z.
By the former cases, (Y,A) and (Y',A') are equioriented and
(A,Z) and (A',Z') are equioriented, hence by Th. 9, (Y,Z) and
(Y',Z') are equioriented.

Because (Y,Z) and (Y',Z') are equioriented for all Y,Z
on s, Y ≠ Z, the congruence of s with t under Γ is an equioriented congruence.   □

Corollary 10

If there exist distinct points A, B on s such that (A,B)
and (A',B') are equioriented, then the congruence of t with s,
under Γ^{-1}, is an equioriented congruence.   (Ex.)

We now have the background we need for the following
Hjelmslev theorem.*

Theorem 11

If X' = X Γ is a motion of H^2 that maps a line s onto
a non-intersecting line t, and if the congruence of s with t,
under Γ , is an equioriented congruence, then the midpoints of
all the segments S[XX'] , X ∈ s, are collinear. Each two such
midpoints are distinct, and their line lies in the strip between
s and t and separates s and t.

Proof

If A and B are any two points of s, the equiorientation

_____

* A more general version of the theorem is stated in the
exercises.

of (A,B) and (A',B') implies that B and B' are in one side
of L(AA'), hence the midpoints of S[AA'] and S[BB'] are
distinct. Now let A,B, and C denote any three points of s,
with Γ-images A', B', C' respectively. Since one of the
points A,B,C must be between the other two, we may suppose
that <ABC>, hence that <A'B'C'>. Let P, Q, R be the midpoints
of S[AA'], S[BB'], S[CC'] respectively, and let the reflection
in point P be represented by $X* = X\Gamma_p$. Since Γ_p is a motion,
<ABC> implies <A*B*C*>, and since $A* = A\Gamma_p = A'$, the line
L(AA') is invariant under Γ_p. The reflection Γ_p interchanges
the sides of L(AA'), hence L(AA') separates R(AB) and R(A*B*) =
R(A'B*) and therefore separates R(A'B*) and R(A'B'). Because
Γ and Γ_p are both motions, d(A,B) = d(A',B') and d(A,B) =
d(A*,B*) = d(A',B*), and therefore

$$d(A',B') = d(A',B*). \qquad (1)$$

Similarly,

$$d(A',C') = d(A',C*). \qquad (2)$$

Now let u denote the perpendicular bisector of S[B'B*].
If B* ∈ t, then R(A'B') and R(A'B*) are opposite rays and, from
(1) and (2) it follows that A' is the midpoint of both S[B'B*]

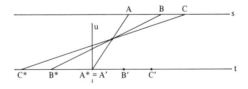

and S[C'C*], thus u is the perpendicular bisector of both
S[B'B*] and S[C'C*]. If B ∉ t then the triangles, ΔB'A'B*
and ΔC'A'C*, exist and, from (1) and (2), both are isosceles

with bases S[B'B*] and S[C'C*] respectively. Since <A'B'C'>
and <A'B*C*>, ∢B'A'B* = ∢C'A'C*. By Th. 4, II-3, the line
u is the bisector of ∢B'A'B* and is therefore the bisector
of ∢C'A'C* and so u, again by Th. 4, II-3, is the perpendi-

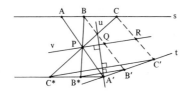

cular bisector of S[C'C*].

Consider the line v that passes through P and is perpendicular to u. The line v bisects S[BB*] and is perpendicular
to u, the perpendicular bisector of S[B*B'], hence by Cor. 12,
III-3, v must bisect S[BB'], the third side of ΔBB*B', and
therefore Q is a point of v. The line v also bisects S[CC*]
and is perpendicular to u, the perpendicular bisector of S[C*C'],
hence by Cor. 12, III-3, v must bisect S[CC'], the third side
of ΔCC*C', and therefore R is a point of v.

If s denotes the set of all midpoints of segments
S[XX'], X ∈ s, we have shown that each three points in s are
collinear. But that implies that s is a linear set, hence
s ⊂ L(PQ) = v.

Assume that v intersects s at a point Y. Then, as just
proved, the midpoints M of S[YY'] is on v. Because s ∩ t = ∅,
Y' ∉ s, so Y' ≠ Y, hence M ≠ Y and therefore v = L(YM) =
L(YY'). For any point W ≠ Y, (Y,W) and (Y',W') are equioriented,
so W and W' are in one side of L(YY'), which implies that S[WW']

and L(YY') = v do not intersect, so the midpoint of S[WW']
is not on v. The contradiction shows that v cannot intersect
s. Similarly, v does not intersect t. Because P ∈ v, and
<APA'>, v must lie in the A'-side of s and in the A-side of
t, so v lies in the strip between s and t. The fact that v
separates A and A', together with s ∩ v = ∅, and t ∩ v = ∅,
implies that v separates s and t.   □

Convention

If s and t are non-intersecting lines and if a congruence
of s with t under a motion X' = XΓ is equioriented, the line
bisecting all the segments S[XX'], X ∈ s, will be referred to
as the Hjelmslev line of s and t with respect to Γ.

Corollary 11

If R(AB)|| R(CD), and if S[AB] ≅ S[CD], then the line
bisecting S[AC] and S[BD] is parallel to both R(AB) and R(CD).

Proof

Let the midpoints of S[AC] and S[BD] be denoted by M and
N respectively. Because S[AB] ≅ S[CD], it follows from Cor. 12,

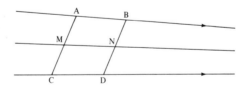

III-5, that there exists a motion Γ such that AΓ = C and BΓ =
D. Since R(AB)|| R(CD) implies that both rays are in the same

side of L(AC), the pairs (A,B) and (C,D) are equioriented.
Thus, by Th. 10, the congruence of s and t, under Γ , is equioriented, and so L(MN) is the Hjelsmlev line of L(AB) and
L(CD) with respect to Γ. Because the line L(MN) intersects
the segment side of the biangle (B-AC-D) and does not intersect either ray side, it is parallel to both ray sides.   □

     If the common perpendicular to two hyperparallel lines s and
t is perpendicular to s at F and to t at G, the midpoint of
S[FG] is the unique center of symmetry to s ∪ t, (c f. Ex.4),
and is called the "center of symmetry to s and t". The following property of such a center will lead us to a simple
construction for the center.

Theorem 12

     If M is the center of symmetry to two hyperparallel
lines s and t, every Hjelmslev line of s and t belongs to the
pencil at M.

Proof

     Let the foot of M in s be denoted by F, and the foot
in t by G, so L(FG) = u is the common perpendicular to s and
t. If v is a Hjelmslev line of s and t then, by definition,
there exists a motion of the plane, say X' = XΓ , such that
sΓ = t and such that the congruence of s with t, under Γ , is
equioriented, and v is the Hjelmslev line of s and t with respect to Γ . If F' = FΓ = G, then because v bisects S[FF'] =
S[FG], v ∈ P(M), so we suppose that F' ≠ G. Because FΓ ≠ G,
then GΓ^{-1} ≠ F. Let GΓ^{-1} be denoted by A, so A' = AΓ = G.
There is a ray R(GB) on t and in the A-side of u and an opposite

open ray R(GC). If F' belonged to R(GB), then R(FF') ⊂
In(∢A'FA) would imply that R(FF') intersected S(AA') at some
point P. In turn, by Th. 5, II-4, <APA'> would imply <FPF'>,
and hence that L(AA') separated F and F'. But the pairs (A,F)
and (A',F') are equioriented, and L(AA') does not separate F
and F'. Thus F' cannot belong to R(GB) and hence must belong
to R(GC).

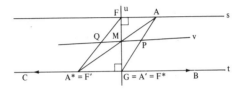

If X* = XΓ_M represents the reflection of the plane in
M, then clearly F* = FΓ_M = A' = G. Since Γ_M interchanges F
and G, it leaves u invariant and interchanges the sides of u.
Thus R[FA] maps onto R[F*A*) = R[GC). Because d(G,A*) =
d(F*,A*)= d(F,A) and d(G,F') = d(A',F') = d(A,F), A* and F'
on R(GC) must be the same point. Therefore AΓ_M = F', and this
with A'Γ_M = F shows that Γ_M maps S[AA'] onto S[F'F]. Thus Γ_M
must map P, the midpoint of S[AA'] onto Q, the midpoint of
S[F'F], and so P, M, and Q are collinear. Because L(PQ) bisects
S[AA'] and S[FF'], L(PQ) is the Hjelmslev line v, hence v ∈ P(M).   □

Theorem 13

Given points A, A₁ and points B,C such that L(AA₁) = s
and L(BC) = t are hyperparallel, then the center of symmetry M
to s and t is constructible, and the line u perpendicular to

both s and t is constructible.

Proof

        The two points on t at distance $d(A,A_1)$ from B are
constructible, (Th. 6), hence the point B_1 on t is constructible such that $d(B,B_1) = d(A,A_1)$, and B_1 is in the A_1-side
of the line $L(AB)$. The points P and P_1 that are the midpoints
of the segments $S[AB]$ and $S[A_1B_1]$ are constructible, by Th. 1,
hence $v_1 = L(PP_1)$ is constructible. A motion Γ_1 exists that
maps A to B and A_1 to B_1. The corresponding pairs (A,A_1) and
(B,B_1) are equioriented, hence by Th. 10, the congruence of s
with t, under Γ_1, is equioriented. Therefore the line v_1
bisecting $S[AB]$ and $S[A_1B_1]$ is the Hjelmslev line of s and t
with respect to Γ_1.

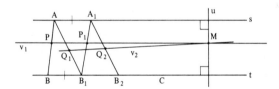

        The point $B_2 = B^{\Gamma_{B_1}}$ is constructible, (Th. 2), and is
a point of t such that $d(B_1,B_2) = d(B_1,B) = d(A,A_1)$. Since
$B_1 \in In(\not\!\!{} BAA_1)$, it is clear that $L(AB_1)$ separates B and A_1.
Also, because $<BB_1B_2>$, $L(AB_1)$ separates B and B_2. Thus A_1
and B_2 are in the same side of $L(AB_1)$. There exists a motion
Γ_2 that maps A onto B_1 and maps A_1 onto B_2. Because the corresponding ordered pairs (A,A_1) and (B_1,B_2) are equioriented,

the congruence of s with t under Γ_2 is equioriented, so a line v_2 that is the Hjelmslev line of s and t with respect to Γ_2 exists. The midpoint Q_1 of $S[AB_1]$ is constructible and the midpoint Q_2 of $S[A_1B_2]$ is constructible. Since v_2 is the bisector of $S[AB_1]$ and $S[A_1B_2]$, $v_2 = L(Q_1Q_2)$ is constructible.

Assume that $v_1 = v_2$. Then $v_1 = L(PQ_1)$ bisects two sides of ΔBAB_1 and therefore, by Th. 12, III-3, v_1 and $L(BB_1) = t$ are hyperparallel, and the perpendicular bisector of $S[BB_1]$ is their common perpendicular. But $v_1 = L(P_1Q_2)$ also bisects two sides of ΔB_1AB_2. Therefore, by Th. 12, III-3, the perpendicular bisectors of $S[B_1B_2]$ is the common perpendicular to v_1 and $L(B_1B_2) = t$. Since the existence of two common perpendiculars to v_1 and t contradicts Th. 13, III-3, the assumption that $v_1 = v_2$ cannot hold. Thus $v_1 \neq v_2$, and by Th. 12, v_1 and v_2 intersect at M, hence M is constructible. By Th. 3, the line through M and perpendicular to s is constructible and this is the line u that is perpendicular to both s and t. □

We turn now to a second of our fundamental constructions, that of the fan angle of a point and line. To obtain this construction, we need a theorem whose proof can be obtained most clearly, we believe, through the steps in a developmental discussion. To this end, we start with a point P that has foot F in a line t, $P \neq F$, and with D the midpoint of $S[PF]$. There

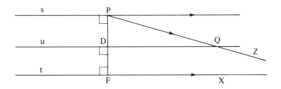

exist lines s and u perpendicular to L(PF) at P and D respectively, and u is an axis of symmetry to the hyperparallels s
and t.   Let an arm of the ⊀(P,t) be denoted by R⟨PZ), with
R(PZ) ‖ R(FX).   Because u)(t, u intersects the biangle (X-PF-
X) at D and at a second point Q, and we may suppose that
< PQZ⟩ .

On line s there is a ray R(PY) in the X-side of L(PF),
and the reflection in u maps R⟨FX) onto R⟨PY) and maps R⟨PZ)
onto a ray R⟨FV) that is parallel to R⟨PY) and that passes
through Q.   We may suppose that < FQV>, hence that ⊀PQF and
⊀ZQV are opposite angles, both bisected by u.   If α = ½⊀ZQV°,

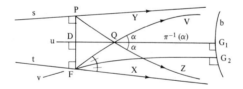

there is a point G_1 on the ray opposite to R(QD) such that
$d(Q,G_1) = \pi^{-1}(\alpha)$, and by Cor. 3.1, III-4, the line b perpendicular to u at G_1 is parallel to both R(QZ) and R(QV).   Because
⊀ZQV is the fan angle of Q and b, b belongs to both the parallel families F[R(QZ)] and F[R(QV)].   Since <PQZ>, R(FX) ‖
R(PZ) => R(FX) ‖R(QZ) => R(FX) ‖ b.   Similarly, because <FQV>,
R(QV) ‖ b => R(FV)‖ b.   Thus ⊀VFX is the fan angle of F and b,
hence the line v that bisects ⊀VFX is perpendicular to b at
a point G_2.

If A is any point on the ray R(FX), A is not on b
because t ∩ b = ∅, and so the equidistant curve EC(b;A) exists.
What is not obvious is that EC(b;A) intersects the ray R(PZ)
at a point B such that d(P,B) = d(F,A) and L(AB) ⊥ s. It is
this perpendicularity that we are most interested in since it
will lead us to a construction for R[PZ).

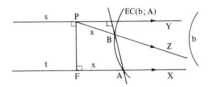

Let A be a point of R(FX) and x = d(F,A). On R(PZ)
there is a point B such that d(P,Z) = x, and on R(FV) there
is a point C such that d(F,C) = x. From x = d(F,A) = d(F,C),
it follows that the triangle ΔAFC is isosceles. Therefore the
line v that bisects ⫨XFV = ⫨AFC is perpendicular to S[AC] at
its midpoint 0. Now, let a = d(P,Q) = d(F,Q), and consider
S[BC]. If x < a, then <PBQ> and <FCQ>, and d(Q,B) = a-x = d(Q,C).

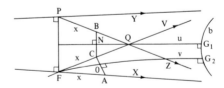

If x > a, then <PQB> and <FQC>, and d(Q,B) = a+x = d(Q,C).
Thus if x ≠ a, ΔBQC is isosceles and N, the midpoint of S[BC]

exists. Because u bisects the opposite angles, ∡PQF and ∡ZQV,
u bisects ∡BQC and is therefore perpendicular to S[BC] at N.

If x ≠ a, the points A, B, C cannot be collinear for
then the quadrilateral OG_2G_1N would have four right angles.
Thus ΔABC exists, and because the perpendicular bisectors
of sides S[AC] and S[BC] are perpendicular to b, it follows from
Th. 6, III-5, that the perpendicular bisector of S[AB] is also
perpendicular to b, and EC(b;A) is the circumcycle of ΔABC.
Let w denote the radial line perpendicular to S[AB] at its midpoint M and perpendicular to b at G_3. We note that when x = a,
then B = Q = C, S[AC] = S[AB], and v = w. Thus in all cases
we have the following property:

the line w, perpendicular to S[AB] at its midpoint

M, is also perpendicular to b.   (*)

We next turn our attention to the line L(DM). Since
(Z-PF-X) = (B-PF-A) is a right biangle with d(P,B) = d(F,A),
it follows from Cor. 11 that L(DM), which bisects both S[PF]
and S[BA], is parallel to R(PZ) and R(FX). Because D is the

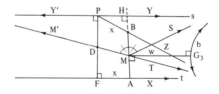

center of symmetry to s and t, the reflection Γ_D maps t onto
s and maps R[FX) in t onto R[PY') opposite to R[PY) in s. The
line L(DM) is invariant and R(DM), parallel to R(FX), maps onto
R(DM'), parallel to R(PY'). Thus if R(MS) denotes the ray at
M parallel to R(PY), then ∡SMD is the fan angle of M and s.

Let the ray at M which is like directed to R(DM) be denoted
by R(MT). From R(DM) || b it follows that R(MT) || b. Also,
R(MS) || R(PY) || b. Thus ⊀TMS is the fan angle of M and b, and
therefore the bisector of ⊀TMS is perpendicular to b. But there
is only one line in the pencil P(M) that is perpendicular to b.
By the (*)-property, w in the pencil P(M) is perpendicular to b,
and so w must be the bisector of ⊀TMS. The two bisectors of
the angles formed by L(TM) and L(SM) are perpendicular, and so
the line perpendicular to w at M must be the bisector of ⊀SMD.
By the (*)-property, L(AB) ⊥ w at M. Therefore L(AB) is the
bisector of ⊀SMD, the fan angle of M and s, hence L(AB) is perpendicular to s at some point H, and we may suppose that <PHY> .

The perpendicularity of L(AB) and s also implies another
property that will be extremely useful later. On line b, let
R(G₃I) be the ray at G₃ which is parallel to R(MS) and R(HY).

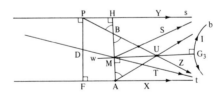

Because L(PB) intersects side S[HM] of the biangle (Y-HM-S), and
does not intersect R[HY), L(PB) must intersect R(MS). Since
L(PB) must intersect two sides of the biangle (S-MG₃-I), and does
not intersect R[G₃I), L(PB) must intersect S(MG₃) at some point
U. The rays R(PZ) and R(BU) are like directed, so R(BU) is
parallel to b in the direction opposite to R(G₃I) on b. The
reflection in line w leaves b and L(AB) invariant, since they
are both perpendicular to w, and interchanges A and B. Therefore

R[AU), the image of R[BU), is also parallel to b, and R(AU)||
R(G_3I). By the transitivity of parallel rays, R(AU)|| R(G_3I)
implies that R(AU)|| R(HY), thus (Y-HA-U) is a right biangle,
and therefore π[d(HA)] = ∢HAU°. Because U is on the perpendicular bisector of S[AB], the triangle ∆ABU is isosceles, hence
∢HAU° = ∢BAU° = ∢ABU°. Since ∢ABU and ∢HPB are opposite angles,
they are congruent, and so ∢ABU° = ∢HBP°. Therefore π[d(HA)] =
∢HBP°.

The following theorem simply summarizes some of the
relations established in the foregoing discussion.

Theorem 14

If (B-PF-A) is a right biangle, with acute angle at P,
and if d(P,B) = d(F,A), then the line s perpendicular to L(PF)
at P is also perpendicular to L(AB) at a point H, and ∢PBH° =
π[d(AH)].

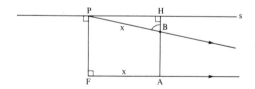

Theorem 15

If a line L(PF) is perpendicular to lines s and t at
points P and F respectively, P ≠ F, and if A on t has foot H
in s, H ≠ P, then the circle C[P,d(F,A)] intersects L(AH)
at two points B and B* such that L(PB) and L(PB*) are the two

lines of the pencil $P(P)$ parallel to t.

Proof

The quadrilateral FAHP is a Lambert quadrilateral with its
acute angle at A. Therefore $d(F,A) > d(P,H) = d[P,L(AH)]$.
Thus, by Cor. 1, II-5, the circle $C[P,d(F,A)]$ intersects $L(AH)$
at two points B and B* such that H is the midpoint of S[BB*].
We suppose the labels chosen so that B is in the t-side of s.

At P there is a ray $R(PZ) \parallel R(FA)$ and on $R(PZ)$ there is
a point C such that $d(P,C) = d(F,A)$. By Th. 14, $L(AC) \perp s$. Also

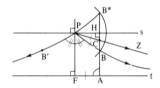

$L(AH) \perp s$, and so $L(AC) = L(AH)$. Suppose that $B \neq C$. Then
$\triangle$ BPC is isosceles, with base S [BC], so the base angles ⊀PBC
and ⊀PCB are both acute. Because B and C belong to R(HA),
either <HCB> or <HBC>. If <HCB>, then ⊀PCB is exterior to $\triangle$PHC,
and $⊀PCB^{o} > ⊀PHC^{o}$ implies that ⊀PCB is obtuse. If <HBC>, then
⊀PBC is exterior to $\triangle$PHB, and $⊀PBC^{o} > ⊀PHB^{o}$ implies that ⊀PBC
is obtuse. In both cases, $B \neq C$ leads to a contradiction, hence
$B = C$, and therefore $R(PB) = R(PC) = R(PZ)$. Thus $L(PB) \parallel t$ in
the direction R(FA) on t. Because s is the perpendicular bisector
of S[BB*] , s bisects ⊀BPB*. Since $L(PF) \perp s$ at P, $L(PF)$ is the
other bisector of the angles formed by $L(PB)$ and $L(PB*)$. Thus

the reflection in L(PF) must map L(PB) onto L(PB*). Because
t is invariant in the reflection, L(PB)|| t implies that
L(PB*)|| t. If B' is the image of B in this reflection, then ∢BPB' is
the fan angle of P and t.   □

Corollary 15.1

Given line t = L(AB) and point P not on t, the fan angle
of P and t is constructible. Thus the open and closed rays at
P and parallel to R(AB) are constructible.   (Ex.)

Corollary 15.2

Given a positive distance a, an angle of measure $\pi(a)$
is constructible.   (Ex.)

Using Th. 13 and Cor. 15.1, we can obtain the third of the
three constructions we set out to find.

Theorem 16

Given an angle ∢APB of measure $\alpha < 90°$, a segment of
length $\pi^{-1}(\alpha)$ is constructible.

Proof

We know that on R(PA) there exists a point F such that
$d(P,F) = \pi^{-1}(\alpha)$, and there exists a line t ⊥ L(PA) at F and
t || R(PB). We will show that t is constructible and hence
that {F}= t ∩ L(PA) is constructible.

We can construct $C = Pr_B$,   (Th. 2), and B is the midpoint of S [PC]. Let u = L(PA). By Th. 3, the points $B' = Br_u$

and C' = CΓ_u are constructible. Because $\angle BPA^o$ = $\angle B'PA^o$.

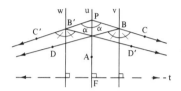

R(PB') || t, and so $\angle BPB'$ is the fan angle of P and the (unknown) line t. By Cor. 15.1, a point D is constructible such that R(BD) || R(PB'). Because <PBC>, R(BC) || t, and since R(BD) || R(PB') || t, $\angle CBD$ is the fan angle of B and t. By Th. 5, the line v bisecting $\angle CBD$ is constructible, and v must be perpendicular to t. The point D' = DΓ_u is constructible, and R(B'D') || R(PB) || t. Since <PB'C'>, R(B'C') || t. Thus $\angle C'B'D'$ is the fan angle of B' and t. The line w bisecting $\angle C'B'D'$ is constructible and is perpendicular to t. Now the lines v and w are hyperparallel, with t as their common perpendicular. Thus, by Th. 13, t is constructible, hence {F} = t ∩ u is constructible, and d(P,F) = $\pi^{-1}(\alpha)$.   □

Corollary 16

Given an angle $\angle BPE$, the line t is constructible such that $\angle BPE$ is the fan angle of P and t.   (Ex.)

It is, of course, apparent that if one uses a physical compass and straight edge to carry out the constructions just described, in an actual drawing, the relations in the diagram will appear to be euclidean. The physically constructed fan angle of P and t will appear to be two oppositely directed

rays, and the drawing of an angle of measure π(a), for a given
length a, will appear to be a right angle. Indeed, if the
physical drawings did not conform to euclidean expectations, the
possibility of a hyperbolic geometry would have been surmised
centuries before its actual discovery.

As a non-mathematical aside, we note that the question
of which mathematical geometry best describes the geometric
character of physical space is a complex problem. One difficulty is deciding what is meant by "points", "lines", etc.
in physical geometry. Also to decide "best" one needs a way
of judging how one conceptual geometry better approximates
physical geometry than another. And however these difficulties
are settled, it is clear from hyperbolic geometry that the
properties of a physical figure cannot safely be judged from
those of a smaller scale model of the figure since the change
in size might affect a change in shape.

Exercises - Section 6

1.   Prove Th. 4.

2.   Prove Th. 5.

3.   Prove Cor. 10.

4.   Let s and t be hyperparallel lines and let u be the line
     perpendicular to s at A and to t at B. If M is the midpoint
     of S[AB], explain why M is a center of symmetry to s ∪ t.
     To show that M is the unique center of symmetry to s and t,
     let P be a point such that the reflection X' = XΓp maps the
     set s ∪ t onto itself. Why must s' = t and t' = s? Why do
     these equalities imply that u' = u? Why must P = M?

5. Prove Cor. 15.1.

6. Prove Cor. 15.2.

7. Prove Cor. 16.

8. Theorem 11 is a special case of the following general Hjelmslev
    property. If each of the sets $\{A_1, A_2, A_3\}$ and $\{B_1, B_2, B_3\}$
    is a triple of distinct, collinear points, and if $d(A_1, A_2) =
    d(B_1, B_2)$, $d(A_2, A_3) = d(B_2, B_3)$, $d(A_3, A_1) = d(B_3, B_1)$, and if
    M_1, M_2, M_3 are such that $A_i r_{M_i} = B_i$, $i = 1,2,3$, then either
    $M_1 = M_2 = M_3$ or else M_1, M_2, M_3 are three distinct, collinear
    points.

9. If $EC(b;P)$ and $EC(b;Q)$ are the two equidistant curves at a
    common distance h from line b, show that the base line b
    bisects $S[PQ]$. If A and B are known to lie on one of the
    curves and C is known to lie on the other, give a construc-
    tion for the base line b.

10. In Ex. 9, if t_P is the tangent line at P and t_Q is the
    tangent line at Q, show that $L(PQ)$, as a transversal of t_p
    and t_Q, forms congruent alternate interior angles with them.

Section 7. Existence Problems; the Method of Associated Right
Triangles

There are some basic questions which we have not yet
answered. For example, given a "triangular set of numbers" a,
b, c (i.e. positive numbers such that the sum of each two is
greater than the third), does there exist a triangle whose sides
have lengths a, b, c respectively? That such a triangle does
exist is a theorem of absolute geometry. In turn, this theorem
implies the basic "two circle" property of absolute geometry,
namely that $C(A_1, r_1)$ and $C(A_2, r_2)$ intersect exactly twice if

r_1, r_2, and $d(A_1, A_2)$ form a triangular set of numbers. What
these theorems indicate is that a thorough treatment of the
foundations of hyperbolic plane geometry requires a more extensive development of absolute plane geometry than was given in
Chapter II. Having noted this, we will not return to such a
development, since this work is only intended to be an introduction to some of the principal features of hyperbolic geometry.
However, in this concluding section we do want to develop the
"method of associated right triangles", which is characteristic
of hyperbolic geometry, and which can be used to settle many
existence questions.

We begin by posing a problem: to what extent can the
measures of the sides and acute angles of a right triangle be
specified? For example, does there exist a right triangle whose
two legs (the sides opposite to the acute angles) have given
lengths? In this case, the answer is clearly "yes". But does
there exist a right triangle whose two acute angles have given
measures? If the sum of the two measures is not less than 90°
we know that the triangle does not exist. However, it is not
obvious, or at least the proof is not obvious, that the triangle
does exist when the sum of the two measures is less than 90°.

To investigate these problems in a systematic way, we
introduce some standardized notation for a right triangle $\triangle ABC$.
We suppose that $\angle C$ is the right angle, that the sides opposite
to A, B, and C have lengths a, b, and c respectively, and that
$\angle A$ and $\angle B$ have measures λ and μ respectively. In terms of these
labels, we list a variety of cases in the following table (the
verification of the first four cases is left to the reader).

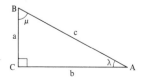

Table I

Case	Given	Restrictions	Rt. Δ Exists
1	two sides a and b	none	yes
2	hypotenuse and side c and a (c and b)	 c > a (c > b)	 yes (yes)
3	hypotenuse and angle c and λ (c and μ)	 λ < 90° (μ < 90°)	 yes (yes)
4	acute angle and side it contains μ and a (λ and b)	 $a < \pi^{-1}(\mu)$ $(b < \pi^{-1}(\lambda))$	 yes (yes)
5	acute angle and opposite side λ and a (μ and b)	 ? (?)	 ? (?)
6	two acute angles λ and μ	λ + μ < 90°	?

We now want to show that the existence of a right triangle implies the existence of four other "associated" right triangles and also the existence of five "associated" L-quadrilaterals.

We will use these associated triangles to obtain the answers
to cases 5 and 6 in Table I from the triangles known to exist
in cases 1 through 4.

In dealing with related measures in the various figures
we will consider, the following notations are useful. Given
the right triangle ΔABC with side lengths a, b, c and angle
measures λ and μ we define

$$\alpha = \pi(a), \beta = \pi(b), \gamma = \pi(c), \ell = \pi^{-1}(\lambda), m = \pi^{-1}(\mu). \qquad (1)$$

Next, we denote complementary angle measures by primes, hence

$$\alpha'=90^{\circ}-\alpha , \beta'=90^{\circ}-\beta, \gamma'=90^{\circ}-\gamma, \lambda'=90^{\circ}-\lambda , \mu'=90^{\circ}-\mu . \qquad (2)$$

Finally, we also use primes to denote complementary distances
of parallelism. That is,

$$a'=\pi^{-1}(\alpha'), b'=\pi^{-1}(\beta'), c'=\pi^{-1}(\gamma'), \ell'=\pi^{-1}(\lambda'), m'=\pi^{-1}(\mu'). \qquad (3)$$

The four numbers a, a', α , α' form a dependent set in the sense
that any one of them determines the other three. Also dependent
in this sense is the set b, b', β , β', the set c, c', γ, γ', the
set $\ell, \ell', \lambda, \lambda'$ and the set m, m', μ, μ'. As a further convention,
it is sometimes convenient to refer to a part in a figure by
its measure. Thus the "a-side" of ΔABC is the side of measure
a and the "λ-angle" is the angle of measure λ.

Now consider the standardized right triangle ΔABC. At
A there exists a ray $R[AX) \perp L(CA)$ and such that $\sphericalangle BAX$ and $\sphericalangle CAB$
are complementary, adjacent angles. Thus $\sphericalangle BAX^{\circ} = \lambda'$. If D
denotes the point on $R(AX)$ at distance $\ell' = \pi^{-1}(\lambda')$ from A, then
the ray $R[DY)$ parallel to $R(AB)$ is perpendicular to $L(AD)$, and
(Y-DA-B) is a right biangle. The line $L(CB)$ intersects the ray
side $R[AB)$ of the biangle, hence must intersect a second side.
Because $L(CB))(L(AD), L(CB) \cap S[AD] = \emptyset$, and so $L(CB)$ intersects $R(DY)$ at some point E, and we may suppose that $<DEY>$.
The parallel construction theorem, (Th. 14, III-6), implies

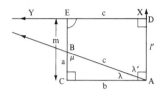

that $d(D,E) = c$ and that $\mu = \sphericalangle CBA^o = \pi[d(C,E)]$, hence $d(C,E) =$
m. Because the definition of the L-quadrilateral CADE began
with $\sphericalangle BAX$ complementary to the λ-angle of the triangle, we call
CADE the λ-related L-quadrilateral associated with $\triangle$ ABC.

We can also determine the measure of the acute angle $\sphericalangle CED$
in the quadrilateral. Let point F on L(CB) be such that <BEF>

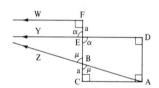

and $d(E,F) = a$. Then $d(B,F) = d(B,E) + a = d(E,C) = m$. If
R(BZ) is like directed to R(AB), $\sphericalangle FBZ$ is the angle opposite to
$\sphericalangle CBA$, hence $\sphericalangle FBZ^o = \sphericalangle CBA^o = \mu$ and $d(B,F) = m = \pi^{-1}(\mu)$. Therefore the ray R[FW] parallel to R(BZ) is perpendicular to L(CF).
Since R(FW) || R(BZ) = > R(FW)| | R(EY), (W-FE-Y) is a right biangle.
Therefore $\sphericalangle FEY^o = \pi[d(F,E)] = \pi(a) = \alpha$, and so $\sphericalangle CED^o = \sphericalangle FEY^o = \alpha$.

The same pattern which defined the λ-related L-quadrilateral associated with $\triangle$ ABC can be used to obtain the μ-related

L-quadrilateral.   At B there is a ray R[BX*) ⊥ L(CB) and such

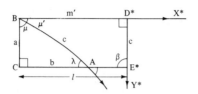

that ∡X*BA° = μ'.   If D* on R(BX*) is at distance m' = $\pi^{-1}(\mu')$
from B, then the ray R[D*Y*) parallel to R(BA) is perpendicular
to L(BD*) and (Y*-D*B-A) is a right biangle.   The line L(CA)
intersects R(D*Y*) at a point E*, and CBD*E* is the μ -related
L-quadrilateral associated with the triangle.   Its successive
sides have measures a, m', c, ℓ, and its acute angle has measure β.

    The parallel construction not only determines two L-quad-
rilaterals associated with a right triangle, it also determines
two right triangles associated with an L-quadrilateral.   Assume
that the L-quadrilateral CADE was given, with acute angle ∡DEC
of measure α containing sides of lengths c and m respectively
and with opposite sides of lengths b and ℓ' respectively.   By
the parallel construction, C(A,c) intersects S(CE) at a point B
such that R(AB) || R(DE), and we call the right triangle Δ ABC
the c-related right triangle associated with CADE.   Also, by
the parallel construction, the circle C(A,m) intersects S(DE)
at a point G such that R(AG) || R(CE) and Δ AGD is the m-related
right triangle associated with the L-quadrilateral.   It is not
difficult to verify that   ∡CBA° = $\pi^{-1}(m)$ = μ and ∡DGA° = $\pi^{-1}(c)$ =
γ , also that ∡CAB° = 90° - π(ℓ') = 90° - λ' =λ and ∡DAB° =

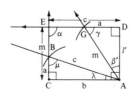

$90^{\circ} - \pi(b) = 90^{\circ} - \beta = \beta'$, and finally that $d(B,C) = d(G,D) = \pi^{-1}(\alpha) = a$.

Starting with either a right triangle or an L-quadrilateral, we can use the associations just described to obtain a sequence of associated right triangles and L-quadrilaterals. For example, starting with the standardized right triangle $\triangle ABC$, there is a λ-related L-quadrilateral CADE whose acute angle contains sides of lengths c and m respectively. The c-related right triangle of this quadrilateral is just the original triangle $\triangle ABC$. But the m-related right triangle is a new triangle (and, in general, not a congruent one), say $\triangle ADG$. The acute angles of $\triangle ADG$ have measures γ and β' respectively. The γ-related L-quadrilateral of $\triangle ADG$ is again the quadrilateral CADE, but the β'-related L-quadrilateral is a new quadrilateral. Using this new quadrilateral, we can obtain a new right triangle, and from that right triangle a new L-quadrilateral, and so on. The sequence closes in the sense that the sixth right triangle in the process is congruent to $\triangle ABC$ and, correspondingly, the sixth L-quadrilateral is congruent to CADE. Two facts about this sequence of figures make the sequence important. First, if one of the figures exists, then all of them exist. Second, if the measures in one of the figures are known, then the measures in all of the figures are known.

Theorem 1

If a right triangle exists whose side lengths and acute angle measures form any one of the sets in the following table, then for each set in the table there exists a right triangle whose measures form that set.   (Ex.)

Table II

Measure set	side one	angle of side one*	hypotenuse	angle of side two	side two
1	a	μ	c	λ	b
2	ℓ'	β'	m	γ	a
3	c'	α'	b'	μ	ℓ
4	m'	λ	a'	β'	c'
5	b	γ	ℓ	α'	m'

(* By the "angle of side one" we mean the acute angle that
contains side one.)

Having devised this rather elaborate system of associated
figures, we now want to show how the system can be used. We
begin by answering the unsettled questions in table I.

Theorem 2

Given any positive number a and any positive angle measure
λ less than 90°, there exists a right triangle with an acute
angle of measure λ opposite to a side of length a.

Proof

The number a and λ determine the sets of numbers a, a',
α, α' and λ, λ',ℓ , ℓ'. In table II, Th. 1, the hypothetical
right triangle with the measure set in row 2 has sides of lengths
ℓ and a respectively. From table I, Case 1, such a right
triangle exists. Designating the ℓ'-side of this triangle as
side one and the a-side as side two, we can define the measure
of the hypotenuse to be m, the angle of side one to have measure
β' and the angle of side two to have measure γ. Thus a right

triangle exists whose measures form set 2 in the table. By
Th. 1 it follows that a right triangle exists whose measure form
set 1 in the table, and in that right triangle an acute angle of
measure λ is opposite to a side of length a.   □

Theorem 3

Given positive angle measure λ and μ such that $\lambda + \mu < 90°$
there exists a right triangle whose acute angles have measure
λ and μ respectively.

Proof

The numbers λ and μ determine the sets of numbers λ,
λ', ℓ, ℓ' and μ, μ', m, m'. In table II, Th. 1, the hypothetical right triangle with the measure set in row 5 has a hypo-
tenuse of length ℓ and a side of length m'. From table I, such
a right triangle exists if $\ell > m'$. Because $\lambda + \mu < 90°$,
$\lambda < 90° - \mu$, that is, $\lambda < \mu'$, which implies that $\pi^{-1}(\lambda)$
$> \pi^{-1}(\mu')$. Therefore $\ell > m'$. Thus a right triangle with hypotenuse
of length ℓ and a side of length m' does exist. Designating
this m'-side as side two, we can define the measure of the angle
of side two to be α', the measure of side one to be b, and the
measure of the angle of side one to be γ. This right triangle
has the measure set 5 in table II, hence by Th. 1 there exists
a right triangle with measure set 1, and in that right triangle
the acute angles have the measures λ and μ respectively.   □

Starting with a right triangle whose measures are known,
one can label the sides of a pentagon in such a way that it
serves as a mnemonic device for the determination of the measures

in the associated right triangles. Suppose that the triangle is
the standardized right triangle Δ ABC that we have been using.
One side of the pentagon is labeled with the length of the hypotenuse, i.e. is labeled c. Next, the two sides of the pentagon
adjacent to the c-side are labeled with the distances of
parallelism of the acute angles in the triangle, i.e. are
labeled ℓ and m.

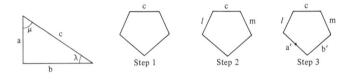

Finally, the two remaining pentagon sides are labeled with the
distances complementary to the lengths of the legs of the triangle, i.e. are labeled a' and b' respectively. However the
choice of the a' and b' sides is important. In the triangle,
the λ-angle is between the c and b sides, but on the pentagon
the corresponding ℓ-side must not be between the c and b' sides.
Similarly, the μ-angle of the triangle is between the c and a
sides, but on the pentagon the m-side must not be between the
c and a' sides. This reversal forces the successive sides of
the pentagon to be in the order c, m, b', a', ℓ. As one can
see from table II, Th. 1, each of these numbers is the length
of a hypotenuse in one of the five associated right triangles.

        To obtain the measures in a right triangle associated
with ΔABC, one simply reasons as though the pentagon had been
labeled starting from that right triangle. For example, there
is an associated right triangle whose hypotenuse has length a'.
For the sides adjacent to a' in the pentagon to have measures

ℓ and b' the acute angles of the triangle have to have measure
λ and β'. Finally, because the remaining pentagon sides are
the c and m sides, the legs of the triangle must have lengths
c' and m'. Because the ℓ-side of the pentagon is between a'
and c sides, the λ-angle of the triangle must <u>not</u> be between
the a'-hypotenuse and the c'side. Thus λ -angle is opposite
to the c' side in the triangle and the β' angle is opposite
to the m' side.

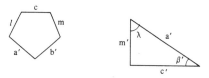

We began this section by posing the problem of the extent
to which the measure in a right triangle could be specified,
and we now have answers to all the cases in Table I. Let us
consider the same problem for S-quadrilaterals and L-quadrilaterals. If ABCD denotes an S-quadrilateral, with M the midpoint
of the base S[AB] and N the midpoint of the summit S[CD], it
is clear that the S-quadrilateral ABCD exists if and only if the
L-quadrilateral AMND exists. Let $d(A,B) = 2a$, $d(A,D) = d(B,C) = m$,

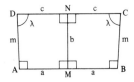

$d(C,D) = 2c$, ∡ACD° = ∡BCD° $= \lambda$, and $d(M,N) = b$. We want to
know to what extent the measures a, m, c, λ , b can be specified.
For example, does an S-quadrilateral exist with given base

length 2a and with summit angles of given measure λ ? Equivalently,

does an L-quadrilateral exist with an acute angle of given measure

λ and with a side, not on this angle, of given length a?

        To answer the questions just posed, we need only consider

the c-related right triangle associated with the L-quadrilateral

AMND, say the right triangle ΔAME, where E is the intersection

of the circle C(M,c) and S(AD). If ΔAME exists then AMND exists

and ABCD exists. In the following theorem about the existence of

S-quadrilaterals, the existence of the corresponding L-quadrila-

terals is not stated but is implied.

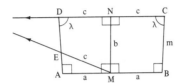

Theorem 4

        If a, m, c, λ , b are positive numbers, a Saccheri quadrila-

teral of base length 2a, summit length 2c, side length m, acute

angle of measure λ, and altitude of length b exists in the following

cases.   (Ex.)

Table III

Case	Given	Restriction	Case	Given	Restriction
1	base, side, 2a, m	none	6	side, summit angle, m, λ	λ < 90°
2	base, summit 2a, 2c	a < c	7	side, altitude m, b	b < m
3	base, summit angle, 2a, λ	λ < 90°	8	summit angle, summit, λ , 2c	λ < 90° c > π⁻¹ (λ)
4	base, altitude 2a, b	β' > π (a)*	9	summit, alt. 2c, b	none
5	side, summit m, 2c	none	10	summit angle, altitude, λ , b	λ < 90°'

---

* (β' = 90° - π(b)).

In congruence-condition theorems, the existence of the
figures is assumed. For example, the side-side-side congruence condition theorem for triangles is independent of the
question about the existence of a triangle with three given
sides. Thus Th. 4 actually gives ten congruence conditions
for S-quadrilaterals (i.e. "base-side", "base-summit", etc.)
and, by implication, ten congruence conditions for L-quadrilaterals

To exemplify the use of associated figures in conjunction with a problem that is interesting and important, we now
want to investigate the problem of determining the tangent lines
to a given cycle c that pass through a given point P exterior to
c. We observe, first, that if L(PA) is such a tangent, with contact point A on c, then L(PA) is perpendicular to the radial line
through A, say the line r_A. Thus L(PA) is not a radial line, and
since P $\notin$ r_A the radial line r_P through P is not r_A. The reflection of the plane in r_P maps c onto itself, maps A to a point A*

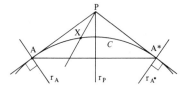

on c, and maps r_A onto the radial line r_{A*}. Because P is fixed,
and the reflection preserves perpendicularity, L(PA*) $\perp$ r_{A*} implies that L(PA*) is also tangent to c. Because L(AA*) is a se-
cant to C, P $\notin$ L(AA*), and so $\measuredangle$APA* exists. By the definition of

tangency, all of c except point A lies in the A*-side of L(PA),
and all of c except A* lies in the A-side of L(PA*).   Thus if
X on c is not A or A*   then X $\in$ In($\angle$APA*) implies that L(PX)
separates A and A* and hence is not tangent to c.   Therefore
if there is a line through P which is tangent to c there are
exactly two such lines.   Moreover, if A and A* are the respective contact points of these tangents, then d(P,A) = d(P,A*),
and r_p is the perpendicular bisector of S[AA*].

       Let us assume, for the moment, that the tangent line
L(PA) exists.   On the radial line r_p there is at least one
point of c (two if c is a circle) and we denote by B the intersection nearest to P.   The line u that is the perpendicular
bisector of S[AB] is a radial line of c, hence the reflection

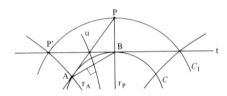

X' = XΓ_u maps c onto itself and maps A to A' = B.   Therefore the
line L(PA), tangent to c at A, must map to L(P'A') = L(P'B) =
t, tangent to c at B.   Let c be the co-radial cycle through P.
Then u is also a radial line of c_1 and therefore P' = PΓ_u must
be on c_1.   Thus P' must be an intersection point of t and c_1.
Also, since Γ_u maps r_p = L(PB) onto L(P'B') = L(P'A) = r_A, $r_{p'}$
= $r_{A'}$ so A is an intersection of $r_{p'}$ and c.

       The last statement above indicates how the existence of
the tangent L(PA) might be proved rather than assumed.   If the line

t, tangent to c at B, does intersect c_1 at some point, call
it P', then the line u that is the perpendicular bisector of
S[PP'] is a radial line of both c_1 and c. Thus the reflection Γ_u maps the cycles onto themselves, so B maps to some
point A on c. The line t, tangent to c at B, must map to a
line t', tangent to c at A. Because P' is on t, its Γ_u-image
must be on t'. But $P = P'\Gamma_u$. Therefore t' = L(PA) is a tangent
that passes through P. Thus the existence of the desired tangents, as well as a method for determining them, will be established if we can show that the line tangent to the cycle c at
B does intersect the co-radial cycle c_1.

If c is a circle, then the fact that B is interior to
the concentric circle c_1 implies that t intersects c_1, by Th.1,
II-5. Thus in the case of a circle, the two tangents through P
do exist. To obtain the same results for a limit circle, we
need the following theorem.

Theorem 5

If line t is perpendicular to the ray R[PB) at B, then
t intersects the limit circle LC[R(PB)] at two points.

Proof

Let L(PX) be the line perpendicular to L(PB) and hence
tangent to LC[R(PB)]. Let a = d(P,B) and let C be a point such
that <PBC>. At B there is a ray R(BY) parallel to R(PX) and,
by Cor. 16, III-6, there exists a line v such that ∢CBY is the
fan angle of B and v.

The biangle (X-PB-Y) is a right biangle, hence $∢PBY^o$ =
$π(a) = α < 90^o$. Therefore t ⊥ L(PB) implies that t subdivides

✦ CBY and hence intersects v at some point Q. On v, let R(QZ) be

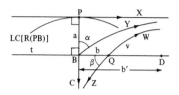

the ray parallel to R(BC) and let R(QW) be the opposite ray which

must be   parallel to R(BY) and hence also parallel to R(PX).

Let b = d(B,Q). Since (C-BQ-Z) is a right biangle, $\angle BQZ^o$ = π(b) = β.

Corresponding to the complementary angle measure β' = 90° - β ,

there is a complementary distance of parallelism b' = π^{-1}(β '),

and on R(BQ) there exists a point D such that d(B,D) = b'. We

will show that D is an intersection point of t and the limit

circle.

        Because Q is interior to the right angle, $\angle$BPX, it foot

F in L(PX) belongs to R(PX), and we may suppose that <PFX>.

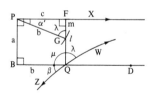

In the L-quadrilateral FPBQ, let c = d(P,F), ℓ = d(F,Q), and

μ = $\angle FQB^o$. The circle C(P,b) intersects S(FQ) at a point G,

and △PFG is the b-related right triangle associated with the

quadrilateral. From this it follows that d(F,G) = π^{-1}(μ) = m,

$\angle FGP^o$ = π^{-1}(ℓ) = λ, and $\angle GPF^o$ = 90° - π(a) = 90°- α = α'.

Define $\pi(c) = \gamma$, $\gamma' = 90° - \gamma$, $c' = \pi^{-1}(\gamma')$, and $\mu' = 90° - \mu$,
$m' = \pi^{-1}(\mu)$. The mnemonic pentagon corresponding to $\triangle PFG$
has its successive sides labeled b, ℓ, c', m', a'. Thus
there exists a right triangle, say $\triangle MNO$, associated with $\triangle PFG$,
whose hypotenuse $S[MO]$ has length c', with legs $S[MN]$ and
$S[NO]$ of lengths a and b' respectively, and with acute angles

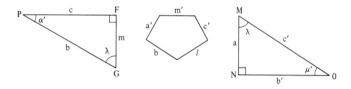

$\angle NMO$ and $\angle NOM$ of measures λ and μ' respectively. By side-
angle-side, $\triangle NMO \cong \triangle BPD$. Thus, by corresponding parts of congruent triangles, $\angle BPD° = \lambda$ and $\angle PDB° = \mu'$.

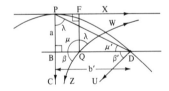

        Now let $R(DU)$ be the ray parallel to $R(PB)$ and hence
also to $R(BC)$. Because $(C-BD-U)$ is a right biangle, and $d(B,D)$
$= b'$, $\angle BDU° = \pi(b') = \beta'$, and $\angle PDU° = \angle PDB° + \angle BDU° = \mu' + \beta'$.
The biangle $(X-FQ-W)$ is also a right biangle, so $d(F,Q) = \ell$
implies that $\angle WQF° = \pi(\ell) = \lambda$. Now, at Q, we have $\angle WQF° + \angle FQB°$
$+ \angle BQZ° = 180°$, hence $\lambda + \mu + \beta = 180°$. This equality implies
that $\lambda = 180° - \mu - \beta = 90° - \mu + 90° - \beta = \mu' + \beta'$. Therefore

$\angle CPD^o = \lambda = \angle PDU^o = \mu' + \beta'$, hence (C-PD-U) is an isosceles
biangle. Thus the perpendicular bisector of S[PD] is para-
llel to R(PC) and R(DU) and is a radial line of the limit circle. Because the reflection in this radial line maps P onto
D, D is an intersection point of t and the limit circle. The
reflection in the radial line L(PB) maps t onto itself and maps
LC[R(PB)] onto itself. It therefore maps D to some point E
which is a second intersection point of t and the limit circle.   □

Corollary 5.1

If <PBC>, the line t tangent to the limit circle LC[R(BC)]
intersects the co-radial limit circle LC[R(PB)] at two
points D and E, and the radial lines through D and E intersect
the inner limit circle at points D* and E* respectively such
that L(PD*) and L(PE*) are tangent to LC[R(BC)].   (Ex.)

Corollary 5.2

If B is interior to a limit circle, every non-radial
line through B intersects the limit circle at two points.   (Ex.)

Equidistant curve properties analogous to the limit circle properties in Th. 5 and its corollaries can be put in the
following form.

Theorem 6

If P has foot F in line s, P ≠ F, and if B is between P
and F, then the line t perpendicular to L(PF) at B intersects
the equidistant curve EC(s;P) at two points.   (Ex.)

Corollary 6.1

If <PBF>, and if line s ⊥ L(PF) at F, the line t tangent to the equidistant curve EC(s;B) at B intersects the
co-radial equidistant curve EC(s;P) at two points D and E,
and the radial lines through D and E intersect the inner equidistant curve at points D* and E* respectively such that L(PD*)
and L(PE*) are tangent to EC(s;B).   (Ex.)

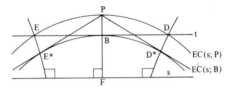

Corollary 6.2

If B is interior to an equidistant curve EC(s;P), and
is in the P-side of s, then every line through B and hyperparallel to s intersects EC(s;P) at two points.   (Ex.)

Since there is no clear cut definition of just how much
work constitutes an introduction to a subject, our decision
to end the chapter at this point is necessarily somewhat arbitrary. However, we feel that the material covered does give a
reasonable representation of both the structure of hyperbolic
plane geometry and its methods. In particular, it constitutes
a background for the next chapter in which we will look at the
subject again, but from a very different point of view.

Exercises - Section 7

1.   Give a justification for the first four cases in Table
      I.

2.   Starting with the standardized right triangle $\triangle ABC$ (used
      for table I), draw and label the parts in the λ-related
      L-quadrilateral, whose acute angle contains an m-side.
      Draw and label the m-related right triangle and show
      that it has measure set 2 in table II and a γ-angle.
      Draw and label the γ-related L-quadrilateral, whose acute
      angle contains a b'-side. Show that the b'-related right
      triangle has measure set 3 in table II and a μ-angle.
      Draw and label the μ-related L-quadrilateral, whose acute
      angle contains an a'-side. Show that the a'-related
      right triangle has measure set 4 in table II and a β'-side.
      Draw and label the β'-related L-quadrilateral, whose acute
      angle contains an ℓ-side. Show that the ℓ-related right
      triangle has measure set 5 in table II and an α'-angle.
      The α'-related L-quadrilateral has an acute angle contain-
      ing a c-side. Show that the c-related right triangle has
      the measure set 1 in table II with which we began. Thus
      verify Th. 1.

3.   Prove Th. 4.

4.   Prove Cor. 5.1.

5.   Prove Cor. 5.2.

6.   Prove Th. 6.

7.   Prove Cor. 6.1.

8.   Prove Cor. 6.2.

9.   In euclidean geometry, corresponding to a circle C(A,a)

there exist six congruent circles $C(B_i, a)$, $i=1,2,\ldots,6$ that form a "ring of circles about $C(A,a)$" in the sense that each of these six circles is tangent to $C(A,a)$ and to two others in the set. Explain why no such configuration can exist in hyperbolic geometry.

10. Show that if n is a positive integer equal to or greater than 7, then there exists a number a and a circle $C(A,a)$ such that there are n circles $C(B_i, a)$, $i=1,2,\ldots,n$ forming a ring about $C(A,a)$ in the sense of Ex. 9.

11. Given $\triangle ABC$ with an angle sum of k°, show that the locus of point X in the C-side of $L(AB)$, and such that the angle sum of $\triangle ABX$ is k°, is an equidistant curve. Clearly k° is an upper bound for the numbers $\angle ABX^\circ$. Give a positive lower bound for these numbers.

Chapter IV. A Euclidean Model of the Hyperbolic Plane

Section 1. An Overview of the Model

At the end of Chapter I, we mentioned the discovery that a non-euclidean geometry could have a euclidean representation. In this chapter, we want to look at one such representation, due to H. Poincaré (1854-1912), which is called "the Poincaré model of hyperbolic geometry". Not only is this model attractively ingenious, but, as we shall explain in detail, it implies that if there is a logical inconsistency in hyperbolic geometry then there is a logical inconsistency in euclidean geometry. Thus, however non-intuitive hyperbolic geometry may appear, it cannot be refuted on logical grounds unless there is a similar refutation of the highly intuitive relations of euclidean geometry.

Throughout this chapter, we suppose that we are dealing with the euclidean plane. We also suppose that standard theorems of elementary euclidean geometry are familiar to the reader and require no justification. Our objective in this section is to define the principal elements in the subsystem of E^2 that constitute the model and to describe what is involved in showing that the model is a representation of hyperbolic geometry. Thus this section is purely introductory in character, an overview of what the model is all about.

The "space" of a Poincaré model is the set of points interior to a euclidean circle. For convenience, we select a unit circle $C(0,1)$ and the set $H = \text{In}[\,C(0,1)\,] = \{\,X: d(0,X) < 1\,\}$ is named the "h-space", and points of H are "h-points". The point 0 is the (euclidean) center of H ; the circle $C(0,1)$, also denoted by C^* , is the (euclidean) boundary to H .

We next want to define certain subsets of H which we will
name "lines". But before we introduce these nominal lines, or
h-lines, we recall some definitions related to euclidean line and
circle intersections. If a line s intersects a circle c at point
P, and is not tangent to c, the angles of intersection of s and
c at P are defined to be the angles formed by s and the line t
that is tangent to c at P. In particular, s is perpendicular, or

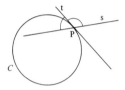

orthogonal, to c, denoted by s ⊥ c, if the angles of intersection
are right angles. Clearly, this occurs if and only if s is a
diameter line of c. If two nontangent circles c_1 and c_2 intersect
at a point P, their angles of intersection at P are defined to be
the angles formed by the lines t_1 and t_2 which are tangent at P to

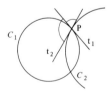

c_1 and c_2 respectively. In particular, c_1 and c_2 are perpendicular,
or orthogonal, at P if the angles of intersection are right angles.

Clearly, c_1 and c_2 are orthogonal at P if and only if t_1 is a

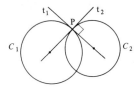

diameter line of c_2 and t_2 is a diameter line of c_1.

We now introduce lines in the h-space.

Definition

(h-line)   A subset of H is an **h-line** if it is the intersection of H with a diameter line of c^* or else is the intersection
of H with a circle orthogonal to c^* .   If L(AB) is a diameter
line of c^* , the h-line t = L(AB) $\cap$ H is a **central** h-line and
L(AB) is the **carrier** of t.   If the circle c is orthogonal to c^* ,
the h-line s = c $\cap$ H is a **non-central** h-line and c is the **carrier**
of s.

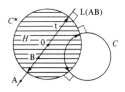

Definition

(Half-planes in H)   If t is a central h-line, with
carrier L(OA), the opposite sides of L(OA) intersect H in the
opposite h-sides, or opposite open h-half-planes, of t.   If t
is non-central, with carrier c , the intersection of the exterior
of c with H is one h-side of t and the intersection of H with the
interior of c is the opposite h-side of t.   The union of t with
a side of t is a closed side of t.

If point A is in H , there is at A a pencil of euclidean
lines $P(A)$ and also a pencil of h-lines $P_h(A)$.   To see why the
model is suggestive of hyperbolic geometry, consider the pencil
$P_h(0)$ and an h-line t whose carrier is orthogonal to c^* at
points B and C.   Let B' denote the point diametrically opposite,
or antipodal, to B and let C' be antipodal to C.   The lines of
$P(0)$ that subdivide the opposite angles, $\not\!\!\xangle BOC$ and $\not\!\!\xangle B'OC'$,

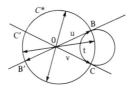

carry the lines of $P_h(0)$ that intersect t.   The h-lines carried
by the lines that subdivide $\not\!\!\xangle BOC'$ and $\not\!\!\xangle COB'$, and those carried
by L(OB) and L(OC), are non-intersectors of t.   Thus $\not\!\!\xangle BOC$ has
the character of the fan angle of 0 and t, with the h-lines u and
v carried by L(OB) and L(OC) respectively, playing the roles of
the parallel s to t through 0.   With the suggestiveness of this
picture in mind, we introduce the following definitions.

Definition

   (Parallel and hyperparallel h-lines)   Two h-lines are
parallel if their carriers intersect at a point of c^*. Two
h-lines are hyperparallel if they do not intersect and are not
parallel. Parallelism of h-lines s and t will be denoted by
s || t and hyperparallelism of s and t by s)(t.

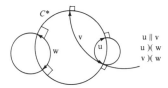

   Though we have defined h-lines in such a way that certain
of their relations are suggestive of the line relations in hyperbolic geometry, the h-space is obviously not as yet a representation of H^2. Because c^* is a unit circle, the euclidean distance
between any two h-points is less than 2. Thus H is a bounded set,
which is certainly not true of H^2 . To correct this aspect of
the model, Poincare introduced a formula for calculating the
distance between two h-points, namely an h-distance, distinct
from the euclidean distance, such that the set of h-distances
for pairs of points in H is an unbounded set. The mathematical
considerations which led him to the particular definition of
h-distance which he adopted are beyond the scope of this book.
A different and elementary derivation of the h-metric is given
in Topic IV of the Appendix. At this point, we simply state the
formula for h-distance. Later, we will verify that it is a
proper distance measure. If A and B are two points of H, they

belong to a unique h-line t, (we haven't yet proved this), whose
carrier intersects C* at some pair of points P*, Q*. The h-distance
between A and B is the number denoted by h(A,B) and defined by

$$h(A,B) = \left| \log \frac{d(P^*,A)d(Q^*,B)}{d(P^*,B)d(Q^*,A)} \right| , \qquad (1)$$

where "d" denotes euclidean distance.*

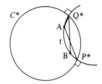

Suppose, for the moment, that formula (1) is a satisfactory
distance measure. Suppose also that we can prove that there
exists a one-to-one correspondence between the points of an h-
line t and the set of all real numbers, i.e. a correspondence
X < - > x, such that Y < - > y and Z < - > z implies that h(Y,Z) =
| y - z | . That is, suppose that we can prove the basic ruler
axiom as a theorem for h-lines. Following the pattern of Sec. 1,
Ch. II, we can define B to be between A and C, in the h-sense,
if A,B,C are distinct points of an h-line and h(A,B) + h(B,C) =
h(A,C). Using this h-betweeness, we can define h-segments and
h-rays as before. And since the proofs for the linear order
properties in Sec. 1, Ch. II, were based on the ruler axiom, the

---

*   One can use a logarithm to any base in (1), so long as that
base is used in defining all h-distances.

corresponding linear order properties in H would follow from the
same proofs.

The last comment is of crucial importance. To show that the
model does represent hyperbolic plane geometry, it is not necessary to prove the h-counterparts of all the theorems in Chapters
II and III. With the h-counterparts of the basic sets established,
together with h-counterparts for the basic relations, one need
only show that the counterparts satisfy the axioms of Ch. II and
Ch. III. Once this is done, each theorem in Ch. II and Ch. III
implies a corresponding h-theorem by the same proof.

In carrying out the program just sketched, there is a
difficulty one can anticipate and which we mention now to
explain the background material of the next section. In absolute
geometry, the side-angle-side congruence condition for triangles
was taken as an axiom. The other triangle congruence conditions
were obtained as theorems, and these theorems were used in the
proofs that point and line reflections in A^2 are motions of A^2
and hence motions of H^2. It is possible to do this in the
reverse order. In fact, if one assumes that the line reflections
of A^2 are motions, one can prove that point reflections are
motions and one can derive the side-angle-side triangle congruence conditions as a theorem. In comparing the Poincaré model
with H^2 , it is this latter procedure we will need to use to obtain
an h-counterpart for the side-angle-side axiom. That is, we will
show that reflections of the h-space in its h-lines do exist and
are motions of H . We will then make use of these motions to
prove that the side-angle-side condition is a congruence condition
for h-triangles.

If t is a central h-line, the reflection of H in t is
obtained from the reflection of E^2 in the euclidean line
carrying t. However, when t is a non-central h-line, the

reflection of H in t is obtained from a mapping of E^2 which
is called a "circular inversion". Since these inversive mappings
play a central role in the study of the model, and since we do
not assume that they are familiar to the reader, they form the
topic of our next section.

Exercises - Section 1

1.   If C(A,a) and C(B,b) are orthogonal at point P, why are
      A, B, and P non-collinear? If t = L(AB), why are the
      circles also orthogonal at Q = P Γ_t?   Why must B be
      exterior to C(A,a) and A be exterior to C(B,b)?

2.   If A is a point of c^*, the boundary circle to H, explain
      how you could construct all the h-lines whose carriers
      pass through A. Thus explain a construction for a "parallel
      family".

3.   If Q is an h-point, explain how you could construct h-lines
      in the pencil $P_h(Q)$.

4.   If L(AB) is a diameter line of c^* that carries an h-line t,
      how would you define the hyperparallel family of h-lines
      with "base line" t?

5.   If t is a central h-line and P is an h-point not on t,
      explain a construction for the two h-lines through P that
      are parallel to t. Is the construction essentially the
      same if t is a non-central h-line?

Section 2. Circular Inversions in E^2

<u>Convention</u>

    If A is a point of E^2, the set of all points distinct
from A form the <u>complement</u> of A, denoted Cp(A). Thus,

$$Cp(A) \; = \; \{X: \; X \neq A \} \; .$$

<u>Definition</u>

    (Circular inversion)   Corresponding to a circle $C(A,r)$,
a mapping defined on Cp(A), and called "the <u>inversion in</u> $C(A,r)$",
is defined as follows: for each $X \neq A$, the image of X is the
point X' on R(AX) such that

$$d(A,X') \; = \; r^2/d(A,X).$$

We will denote the inversion by $\Phi_A(r)$, so $X' = X \Phi_A(r)$.
Point A is the <u>center</u> of the inversion, r is the <u>radius</u> of
inversion, and $C(A,r)$ is the <u>circle of inversion</u>.

    A mapping Γ, which is not the identity, and which has the
property that $Q = P\Gamma$ implies $P = Q\Gamma$ is called an "involution".
Point and line reflections are involutions, as are circular
inversions. If $P \neq A$, R(AP) exists, and corresponding to $C(A,r)$
there is a unique point Q on R(AP) such that $d(A,Q) = r^2/d(A,P)$.
By definition, $Q = P\,\Phi_A(r)$. From $Q \in R(AP)$ it follows that
$P \in R(AQ)$. Also, $d(A,Q) = r^2/d(A,P)$ implies that $d(A,P) =$
$r^2/d(A,Q)$. Therefore $P = Q\,\Phi_A(r)$, hence $\Phi_A(r)$ is an involution.

    The circular inversion $\Phi_A(r)$ leaves each point of
$C(A,r)$ fixed and interchanges points interior and exterior to
$C(A,r)$. If P is on $C(A,r)$, $d(A,P) = r$ implies that $r = r^2/d(A,P) =$
$d(A,P')$, hence $P' = P$. If $P \neq A$ is interior to $C(A,r)$,
$d(A,P) < r$ implies that $r < r^2/d(A,P) = d(A,P')$, hence P'
is exterior to $C(A,r)$. Similarly, if P is exterior to $C(A,r)$,
then $d(A,P') < r$. Since each point of Cp(A) has a unique image
in Cp(A) and is also itself the image of a unique point in

Cp(A), the inversion $\Phi_A(r)$ is a one-to-one mapping of Cp(A)
onto itself. We summarize our conclusions thus far.

Theorem 1.

A circular inversion $\Phi_A(r)$ is a one-to-one, involutory
mapping of Cp(A) onto itself, which leaves each point of C(A,r)
fixed and which interchanges the sets Ex[C(A,r)]   and
In [C(A,r)] ∩ Cp(A)

Our next theorem yields a simple construction for the
inverse of a given point with respect to a given circle.

Theorem 2.

If Q is exterior to C(A,r) and P is interior to C(A,r),
P ≠ A, and if the line perpendicular to L(AP) at P intersects
C(A,r) at a point B such that L(QB) is tangent to C(A,r), then
P and Q are inverse points with respect to C(A,r).

Proof

By hypothesis, ΔAPB and ΔABQ are right triangles,
and since   ∡BAP =   ∡QAB, ΔAPB ~ ΔABQ, by the angle-

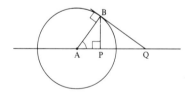

angle similarity condition. Thus, d(A,P)/d(A,B) = d(A,B)/d(A,Q).
This proportion, together with d(A,B) = r, implies that
d(A,P)d(A,Q) = r^2, hence P and Q are inverse points with respect
to C(A,r).   □

Our next theorem plays a key role in determining the action
of an inversion on the lines and circles of E^2.

Theorem 3.

If points B and C are not collinear with the center A of
the inversion X' = X $\Phi_A(r)$, then $\triangle$ ABC ~ $\triangle$AC'B', hence
$\measuredangle$ ABC $\cong$ $\measuredangle$AC'B' and $\measuredangle$ACB $\cong$ $\measuredangle$AB'C'.

Proof.

By the definition of the inversion, $\measuredangle$BAC and $\measuredangle$C'AB'
are the same angle. Also, by definition, $d(A,B)d(A,B') = r^2$
and $d(A,C)d(A,C') = r^2$, and these equalities imply that
$$d(A,B)/d(A,C) = d(A,C')/d(A,B').$$

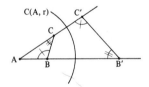

Thus, by the side-angle-side similarity theorem for triangles,
$\triangle$ABC ~ $\triangle$AC'B'.   □
    To see how an inversion acts on lines and circles, first
consider a line t through the center A of the inversion
$\Phi_A(r)$.   If B and C denote the points of intersection of t
with C(A,r), it is clear that R(AB) and R(AC) each maps onto
itself. Thus t $\cap$ Cp(A) is a fixed set in the inversion. With
the exception of the center understood, it is conventional to
say that a line through the center of an inversion maps onto
itself.
    Next suppose that t does not pass through A. Let F be the
foot in t of point A, and F' = F$\Phi_A(r)$. If X on t is not F,

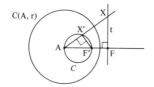

∢AFX is a right angle, and so, by Th. 3, ∢AX'F' is a right
angle. Thus X' lies on the circle c with S[AF'] as diameter.
Conversely, if Y on this circle is neither A or F', then
∢AYF' is inscribed in a semi-circle and hence is a right
angle. From Th. 3, it follows that ∢AFY' is a right angle,
and so Y' ∈ t. Thus, $\Phi_A(r)$ maps t onto c ∩ $C_p(A)$ and
therefore maps c ∩ $C_p(A)$ onto t. It is conventional to say
that an inversion at A interchanges circles through A with lines
not through A. The following is a more precise statement of
our conclusions.

Theorem 4.

If line t passes through A, the inversion X' = X $\Phi_A(r)$
maps the set t ∩ Cp(A) onto itself. If A has foot F in line
t, and F ≠ A, $\Phi_A(r)$ maps t onto the set c ∩ $C_p(A)$, where
c is the circle with diameter S[AF'] . Conversely, if c
is a circle through A, with a diameter S[AB] , $\Phi_A(r)$ maps
the set c ∩ $C_p(A)$ onto the line perpendicular to L(AB) at B'.

Theorem 5.

If a circle C(P,a) is not C(A,r) and does not pass through
A, the inversion X' = X $\Phi_A(r)$ maps C(P,a) onto a circle c'
whose center is collinear with A and P. If S [BC] is the diameter
of C(P,a) on L(AP), then S[B'C'] is the diameter of c' on L(AP).

Proof.

By hypothesis, neither B nor C is A, so there are two possibilities; either A ∉ S[BC] or else A ∈ S(BC). We consider these cases separately.

Case 1

A ∉ S [BC] . Then either <ABC> or <ACB> and we may suppose that <ABC> , which implies that <AC'B' >. If X is any point of C(P,a), distinct from B and C, ⊁BXC° = 90°. Since X' on R(AX) is not on L(AP), and <AC'B' >, C' ∈ In(⊁AX'B').
Therefore,

$$\text{⊁C'X'B'}° = \text{⊁AX'B'}° - \text{⊁AX'C'}°. \tag{1}$$

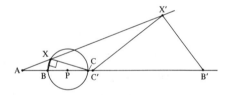

Thus, by Th. 3,

$$\text{⊁C'X'B'}° = \text{⊁ABX}° - \text{⊁ACX}°. \tag{2}$$

Because <ABC> , ⊁ ABX is an exterior angle to Δ BXC, and therefore ⊁ABX = ⊁BXC° + ⊁BCX° = 90° + ⊁ACX°. This equality in combination with (2) implies that ⊁C'X'B'° = 90°. Therefore X' belongs to the circle c' with diameter S[C'B'], By an entirely similar argument, if Y on c' is not B' or C', then ⊁C'YB' is a right angle and this implies that ⊁BY'C is a right angle and hence that Y' belongs to C(P,a). Thus Φ_A (r) interchanges C(P,a) and c'.

Case 2

A ∈ S(BC). (Ex.)   □

Since the non-central h-lines of the Poincaré model are
carried by circles orthogonal to the boundary C*, we have a
special interest in orthogonal circles. In the study of such
circles and how they relate to inversions, the following
concept is extremely useful.

Definition.

(Power of a point with respect to a circle)   Corresponding
to a circle $C(B,b)$, the power of a point P with respect to
$C(B,b)$ is the number

$$\text{Pw}[P;C(B,b)] = d(P,B)^2 - b^2$$

Theorem 6.

A point is interior to the circle $C(B,b)$ if and only if
its power with respect to the circle is negative. The center
B, whose power with respect to $C(B,b)$ is $-b^2$, is the unique
point of least power. Points of the plane have zero power if
and only if they belong to $C(B,b)$ and have positive power if
and only if they are exterior to $C(B,b)$.   (Ex.)

Theorem 7.

A circle $C(P,a)$ is orthogonal to the circle $C(B,b)$ if and
only if the power of P with respect to $C(B,b)$ is a^2.

Proof

Suppose first that $C(P,a)$ is a circle orthogonal to $C(B,b)$
at a point D. The definition of orthogonality implies that
$\triangle PDB$ is a right triangle with hypotenuse $[PB]$. From the
Pythagorean theorem, $d(P,B)^2 - d(B,D)^2 = d(P,D)^2$, and therefore
$d(P,B)^2 - b^2 = a^2 = \text{Pw}[P;C(B,b)]$.

Conversely, suppose that C(P,a) is a circle such that the power of P with respect to C(B,b) is a^2. Then a is a positive number, and $a^2 > 0$ implies that P is exterior to C(B,b). Thus there exists a line L(PT) tangent to C(B,b) at T, and the circle C(P,d(P,T)) is orthogonal to C(B,b) at T. By the argument of the previous paragraph, the power of P with respect to C(B,b) must be $d(P,T)^2$. Therefore $d(P,T)^2 = a^2$ and, since a and d(P,T) are positive, $d(P,T)^2 = a^2$ implies that d(P,T) = a. Therefore C(P,a) is the circle C(P,d(P,T)) and hence is orthogonal to C(B,b).   □

Corollary 7.

If L(PT) is tangent to C(B,b) at T, the power of P with respect to C(B,b) is $d(P,T)^2$.

Two familiar theorems of euclidean geometry provide a different way of calculating the power of a point.

Theorem 8.

If P is interior to the circle C(B,b) and one secant through P intersects the circle at C and D, and a second secant through P intersects the circle at E and F, then d(P,C)d(P,D) = d(P,E)d(P,F).   (Ex.)

Corollary 8.

The power of P with respect to C(B,b) is -d(P,C)d(P,D).

Proof

If P = B, the power of P with respect to C(B,b) is $-b^2$

and this is also -d(P,C)d(P,D). If P ≠ B, then L(PB)

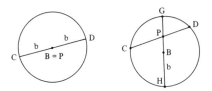

intersects C(B,b) at points G and H and we may label these so
that   < GPB >   and   < PBH > .   From Th. 8, d(P,C)d(P,D) =
d(P,G)d(P,H).   Since d(P,G) = b - d(P,B) and d(P,H) = b + d(P,B),
it follows that d(P,C)d(P,D) = b^2 - d(P,B)^2. Thus -d(P,C)d(P,D) =
d(P,B)^2 - b^2 = Pw [P;C(B,b)] .   □

Theorem 9.

If P is exterior to the circle C(B,b) and one secant through
P intersects the circle at C and D, and a second secant through
P intersects the circle at E and F, then d(P,C)d(P,D) = d(P,E)d(P,F).
(Ex.)

Corollary 9.

The power of P with respect to C(B,b) is d(P,C)d(P,D).

Proof

Let the intersections of L(P,B) with C(B,b) be labeled G
and H in such a way that   <PGB>   and   <GBH> .   By Th. 9,

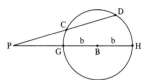

$$d(P,C)d(P,D) = d(P,G)d(P,H) = [d(P,B)-b][d,(P,B) + b] =$$
$$d(P,B)^2 - b^2 = Pw[P;C(B,b)]. \qquad \square$$

Our next two theorems will be especially useful when we return to the study of the Poincaré model.

<u>Theorem 10</u>.

If P and Q are inverse points with respect to a circle C(A,r) then every circle that passes through P and 0 is orthogonal to C(A,r).

<u>Proof</u>

Let c denote a circle that passes through P and Q. Because P and Q are inverses with respect to C(A,r), A does not belong to S[PQ] and hence, by Th. 1, II-5, A is exterior to c.

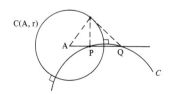

From Cor. 9 it follows that the power of A with respect to c is $d(A,P)d(A,Q)$. But $d(A,P)d(A,Q) = r^2$. Therefore, by Th. 7, $C(A,r)$ is orthogonal to c.   □

Theorem 11.

A circle is invariant under the inversion $\Phi_A(r)$ if and only if it is $C(A,r)$ or is orthogonal to $C(A,r)$.

Proof

Let c denote a circle that is orthogonal to $C(A,r)$ at points B and D. Since A is exterior to c, it is not on c, hence c inverts to a circle c'. The points B and D on c are fixed points of the inversion, so they also belong to c'. The lines L(AB) and L(AD) are the only two tangents to c that pass through A. Thus if X on c is neither B or D, then L(AX) is a secant to c and intersects c at a second point Y. By Cor. 9, the power of A with respect to c is $d(A,X)d(A,Y)$. But, by Cor. 7, the power of A with respect to c is $d(P,B)^2 = r^2$. Therefore $d(A,X)d(A,Y) = r^2$, so X and Y are inverses and each is the image of the other in the inversion. Thus, except for the fixed points B and D, the inversion $\Phi_A(r)$ interchanges points of c in pairs, and $c' = c$.

Next, suppose that a circle c, distinct from $C(A,r)$, inverts onto itself. Because c can intersect $C(A,r)$ at no more than two points, c has a non-fixed point X which inverts to a point X' on c. By the definition of inversion, A on L(XX') is not on S[XX'] and is therefore exterior to c. Thus, by Cor. 9, the power of A with respect to c is $d(A,X)d(A,X')$. Since $d(A,X)d(A,X') = r^2$, it follows from Th. 7 that $C(A,r)$ is orthogonal to c.   □

Corollary 11.

    If circles c_1 and c_2 are orthogonal, then every line through the center of c_1 that is a secant to c_2 intersects c_2 at a pair of points which are inverses with respect to c_1.

Exercises - Section 2

1.   In an inversion $X' = X\Phi_A(r)$, why does $<ABC>$ imply that $<AC'B'>$?

2.   Use Th. 2 to explain a construction for the point $P\Phi_A(r)$: (i) when $P \in In [C(A,r)]$; (ii) when $P \in Ex [C(A,r)]$.

3.   In the inversion $\Phi_A(r)$, if $<ABC>$ and if $d(A,B) = 2$ and $d(A,C) = 4$, compute $d(A,B')$ and $d(A,C')$, where B', C' are the images of B and C respectively. By Th. 5, the circle with diameter $S[BC]$ inverts to the circle with diameter $S[B'C']$. Does the center of the first circle invert to the center of the second? Explain.

4.   How could you choose the radius r in Ex. 3 so the circle with diameter $S[BC]$ maps onto itself?

5.   Give the proof for Case 2 in the argument for Th. 5.

6.   Prove Th. 6.

7.   Explain Corollary 7.

8.   Prove Th. 8. (Show that $\triangle PCF \sim \triangle PED$).

9.   Prove Th. 9.

10. In the inversion $\Phi_A(r)$, what circles invert to the lines tangent to $C(A,r)$?

11. If A,P, and Q are non-collinear, prove that there is exactly one circle through P and Q that is orthogonal to $C(A,r)$.

Section 3. Angle and Cross Ratio Invariance Under Inversion

In the last section, we established which lines and circles
are invariant in an inversion. In this section we will establish
the invariance of two measures which play an important role in
the structure of the Poincare model. The first of these invariants concerns the angles at which curves intersect, and the
second concerns the ratio of distances involved in the h-metric
of the model, and we will consider them in that order.

A mapping Γ that preserves the tangency of curves and
preserves the angles at which curves intersect is a <u>conformal</u>
<u>mapping</u>. Though circular inversions are conformal mappings, to
prove this with complete generality requires the definition of
general curves and their tangents. These definitions involve
analytic methods not in keeping with the elementary character of
the present text. Fortunately, it is sufficient for our
purposes to establish a less general result, namely that inversions act conformally on the lines and circles of E^2.* That
is, we will show that if s_1 and s_2 are two lines, two
circles, or a line and a circle, intersecting at a point P,
then under inversion the image sets s'_1 and s'_2 intersect
at a point P' in such a way that the angles at which they
intersect have the same measures as the angles at which s_1
and s_2 intersect. We begin with some names and notations that
will be convenient.

<u>Conventions</u>:

The collection of all lines and circles which pass through
a point P, the <u>line-circle</u> <u>family</u> at P, will be denoted by F_p,
and a member of F_p will be referred to as a "curve" whether
it is a line or a circle. If s is a curve in F_p, the curve
s together with all curves in F_p which are tangent to s at P

---

* The proof we will give requires only slight modification to
be completely general once general tangency has been defined.

will be denoted by $F_p(s)$.

Clearly, for each $s \in F_p$, the subfamily $F_p(s)$ contains exactly one line. Also, if s_1 is a curve in F_p which is tangent to s at P, then $F_p(s) = F_p(s_1)$.

Theorem 1.

If points P and P' are inverses with respect to the circle $C(A,r)$, the inversion $\Phi_A(r)$ maps the line-circle family F_p onto the line-circle family $F_{p'}$, and two curves in F_p which are tangent at P have images which are tangent at P'.

Proof

If s is a curve in F_p and s' is its image under $\Phi_A(r)$, then $P \in s$ implies that $P' \in s'$. By definition, s is either a line or a circle and so, by Th. 4 and Th. 5, IV-2, s, is either a line or a circle. Thus $s' \in F_{p'}$, and F_p maps into $F_{p'}$. But if R is a curve in $F_{p'}$, then, as just shown, R inverts to a curve R' in F_p. Since R' inverts to R, each curve in $F_{p'}$, is the image of curve in F_p, and so F_p maps onto $F_{p'}$.

Now let s_1 and s_2 denote two curves in F_p which are tangent at P. Then s_1 and s_2 cannot be two lines. Whether they are two circles or a line and a circle, their tangency at P implies that they intersect only at P. Therefore their respective images s'_1 and s'_2 intersect only at P'. If s'_1 and s'_2 were both lines, then, by Th. 4, IV-2, s_1 and s_2 would both be circles in F_A, hence would have A and P as two intersections, contradicting their tangency. Thus s'_1 and s'_2 must be two circles or a line and a circle. In either case, since they intersect only at P' they are tangent at P'.   □

Corollary 1.1

For each $s \in F_p$, the family $F_p(s)$ maps onto the family $F_{p'}(s')$.   (Ex.)

Corollary 1.2

If s_1 and s_2 in F_p are not tangent at P, then their images s'_1 and s'_2 are not tangent at P'.   (Ex.)

If s_1 and s_2 are two non-tangent curves in F_p, the line t_1 in $F_p(s_1)$ and the line t_2 in $F_p(s_2)$ form the angles of inter-section of s_1 and s_2. In an inversion which maps P to P', s_1 and s_2 have non-tangent images s'_1 and s'_2 respectively and the line $v_1 \in F_{p'}(s'_1)$ and the line $v_2 \in F_{p'}(s'_2)$ form the intersection angles of s'_1 and s'_2 at P'. We want to prove not only that the angles formed by t_1 and t_2 have the same measures as those formed by v_1 and v_2, but also that the inversion pairs each individual angle of intersection at P with a particular and congruent angle of intersection at P'. To define the corres-pondence of individual angles of intersection in an inversion, we need the notion of the tangency of a ray and a circular arc, so we review some basic arc concepts.

Since we will not consider arcs on curves other than circles, the term "arc" will always refer to a circular arc. If P and Q are two points of a circle c, the intersection of c with a side of L(PQ) = t is an open arc of P and Q on c , denoted by arc_1(PQ) or by arc(PAQ), if A belongs to the open arc. The intersection of c with the opposite side of t is the opposite open arc, denoted by arc_2(PQ) or by arc(PBQ) if B belongs to the arc. Points P and Q are the endpoints to the opposite arcs, and the union of P and Q with arc_1(PQ) is a closed arc denoted by arc_1[PQ]. If S[PQ] is a diameter of c , the opposite arcs of P and Q on C are opposite <u>semicircles</u>. If S[PQ] is not

a diameter of C , and if E is the center of C , the (open or

closed) arc of P and Q which intersects the E-side of t is the

<u>major arc</u> of P and Q and the opposite (open or closed) arc is

the <u>minor arc</u> of P and Q.

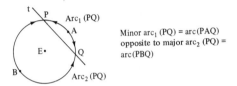

Minor arc$_1$ (PQ) = arc(PAQ)
opposite to major arc$_2$ (PQ) =
arc(PBQ)

Definition

(Tangent ray and arc)   A ray R is <u>tangent to a circular</u>

<u>arc</u> S at point A if: (i) A is the origin to R and an endpoint

to S ; (ii) the line of R is tangent at A to the circle of S ;

(iii) the open ray of R and the open arc of S lie in the same

side of the line through the endpoints to S .

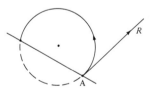

It should be noted that the tangency just defined does not

depend upon the openess or closedness of either the ray or arc.

If R(AX) is tangent to arc(AB), then both R(AX) and R [AX] are

tangent to all four of the arcs of A and B which contain arc(AB).

The following properties of arc-ray tangency are easily verified.

Theorem 2.

If arc$_1$(AB) and arc$_2$(AB) are opposite arcs, there is

exactly one open ray R(AX) which is tangent to arc_1(AB), and
the opposite open ray is tangent at A to arc_2(AB).   (Ex.)

Theorem 3.

   If L(AX) is tangent to a circle c at A and point B on c
is distinct from A, then exactly one of the opposite arcs
arc_1(AB) and arc_2(AB) is tangent to R(AX).   (Ex.)

Theorem 4.

   If R(AX) is tangent to arc(AB) then arc(AB) is interior
to ⊀XAB.   (Ex.)

Theorem 5.

   If R(AX) is tangent to arc(AB) on circle c , then arc(AD)
on c is tangent to R(AX) if and only if arc(AB) $\subset$ arc(AD) or
else arc(AD) $\subset$ arc(AB).   (Ex.)

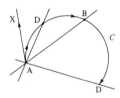

Conventions:

   The collection of all arcs which are tangent to a ray is
the arc family of the ray. The arc families of two opposite
rays are opposite arc families.

   We now introduce the notion of corresponding rays in an
inversion, and this will give us a natural way of defining
corresponding intersection angles.

Definition

(Corresponding rays in an inversion)   Rays R_1 and R_2 are corresponding rays in an inversion $\Phi_A(r)$ if their origins are inverses with respect to $C(A,r)$ and there exists an arc tangent to R_1 and an arc tangent to R_2 which are inversive images of each other.

We note that the definition of corresponding rays is such that if rays R_1 and R_2 correspond then the open and closed rays of R_1 correspond with R_2 and each corresponds with the open and closed rays of R_2. Also, Th. 1 and Th. 2 imply the following property.

Theorem 6.

If rays R_1 and R_2 correspond in an inversion, then a ray opposite to R_1 corresponds with a ray opposite to R_2.   (Ex.)

Theorem 7.

If B is not the center of the inversion $\Phi_A(r)$, then each ray at B has a unique corresponding open ray in the inversion.

Proof

Let primes denote images under $\Phi_A(r)$, with $B' = B^{\Phi_A(r)}$. We first consider a ray R(BX) which is not on L(AB). Let s denote a circle which is tangent to L(BX) at B and which does not pass through A. Because L(AB) is not L(AX), it is not tangent to s and hence intersects s at B and at a second point C. The circle s maps to a circle s'. By Th. 3, there exists $arc_1(BC)$ on s which is tangent to R(BX). The image of $arc_1(BC)$

is $\text{arc}_1(B'C')$ on s', and, by Th. 2, there exists a ray $R(B'Y)$
which is tangent to $\text{arc}_1(B'C')$ and is therefore a ray corresponding to $R(BX)$. The inversion $\Phi_A(r)$ maps $L(AB)$ onto

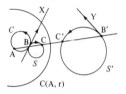

$C(A, r)$

itself and leaves each side of $L(AB)$ invariant. Therefore
$\text{arc}_1(BC) \subset$ X-side of $L(BC) =$ X-side of $L(AB)$ implies that
$\text{arc}_1(B'C') \subset$ X-side of $L(AB)$, hence that $R(B'Y)$ is in the
X-side of $L(AB)$. Since $L(BX) \neq L(AB)$, there exists a unique
circle c which passes through A and is tangent to $L(BX)$ at B.
This circle inverts to a line through B' and because $c \in F_B(s)$
it follows that $c \; \Phi_A(r) \; \in \; F_{B'}(s')$, hence that $c \; \Phi_A(r)$
is the line $L(B'Y)$. Thus $\text{arc}_1(BA)$ on c and in the X-side of
$L(AB)$ inverts to $R(B'Y)$.

     Now suppose that $R(B'Z)$ is a ray which corresponds to
$R(BX)$. Then, by hypothesis, there exists $\text{arc}_1(BD)$ tangent to
$R(BX)$ and such that its image $\text{arc}_1(B'D')$ is tangent to $R(B'Z)$.
Since $L(AB)$ is not $L(BX)$, $L(AB)$ is not tangent to the circle
s_1 of $\text{arc}_1(BD)$ and so it intersects s_1 at B and at a second
point c_1. On s_1, one of the opposite arcs of B and c_1,
say $\text{arc}_1(BC_1)$, is tangent to $R(BX)$ and therefore, by Th. 5,
$\text{arc}_1(BC_1) \subset \text{arc}_1(BD)$ or $\text{arc}_1(BD) \subset \text{arc}_1(BC_1)$. Correspondingly,
$\text{arc}_1(B'C_1') \subset \text{arc}_1(B'D')$ or $\text{arc}_1(B'D') \subset \text{arc}_1(B'C_1')$. In either
case, it follows that $\text{arc}_1(B'C_1')$ is tangent to $R(B'Z)$. But now,
by exactly the same argument as before, $R(B'Z)$ must be the image
of $\text{arc}_1(BA)$ on c, and therefore $R(B'Z) = R(B'Y)$. Thus the

open ray corresponding to R(BX) is unique.

Next, suppose that R(BX) is on L(AB), say R(BX) = R(BA).
Let s denote a circle which is tangent to L(AB) at B. The
circle c_1, with S [AB] as diameter, is orthogonal to s and
intersects s at B and at a point E. Let arc_1(BE) denote the arc
of B and E on s that is interior to c_1. Under the inversion
$\Phi_A(r)$, s maps to a circle s' which is tangent to L(AB) at
B' and c_1 inverts to the line u which is perpendicular to
L(AB) at B'. The interior of c_1 maps to the non-A-side of u.

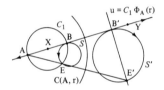

so the image of arc_1(BE) is the semicircle arc_1(B'E') on s'
and in the non-A-side of u. Thus the ray R(B'Y) which is
tangent to arc_1(B'E'), and corresponds   to R(BX), must be in
the non-A-side of u. Since L(B'Y) is perpendicular to u,
R(B'Y) is the open ray at B' on L(AB) which is oppositely
directed to R(BX).

If R(B'Z) is a ray which corresponds to R(BX) = R(BA) then,
by hypothesis, there exists arc_1(BF) tangent to R(BA) and such
that its image arc_1(B'F') is tangent to R(B'Z). The circle c_1
is orthogonal to the circle s_1 of arc_1(BF) and intersects it at
B and at a second point E_1. One of the arcs of B and E_1 on s_1,

say $arc_1(BE_1)$, is tangent to R(BA) and is therefore, by Th. 4,
interior to C_1 . Now, as argued before, it follows that the
image of $arc_1(BE_1)$ is tangent to R(B'Z) and that R(B'Z) must be
the ray at B' on L(AB) which is oppositely directed to R(BA).
Thus R(B'Z) = R(B'Y) and so the ray corresponding to R(BX) is
unique. If $R(BA_1)$ is the open ray opposite to R(BA), then the
open ray corresponding to $R(BA_1)$ must be the open ray opposite
to R(B'Y) and is therefore uniquely determined.   □

Corollary 7.1 If B' is the image of B in an inversion $\Phi_A(r)$,
then R(BX) on L(AB) corresponds with an oppositely directed
ray on L(AB) at B'. A ray R(BX) which is not on L(AB) lies in
the same side of L(AB) as one of the open arcs of A and B on
the circle through A and tangent to L(BX) at B, and the open
ray corresponding to R(BX) is the image of this arc.

Corollary 7.2 If rays R_1 and R_2 correspond in an inversion,
then an arc which is tangent to R_1 and not on a circle through
A inverts to an arc which is tangent to R_2.

Definition. (Corresponding angles in an inversion) Two angles
are corresponding angles in an inversion if the arms of one
are the two closed rays which correspond respectively with
the arms of the other.

Corollary 7.3 If B' is the image of B in an inversion, then
each angle at B has a unique corresponding angle at B', and
if ⦠XBY and ⦠UB'V correspond in the inversion, then their
opposite angles correspond in the inversion.   (Ex.)

__Theorem 8.__   If an inversion $\Phi_A(r)$ maps two non-tangent curves
s_1 and s_2 in a family F_B to curves s_1' and s_2' respectively in
$F_{B'}$, then the four angles which correspond to the intersection
angles of s_1 and s_2 at B are the four intersection angles of s_1'
and s_2' at B'.

__Proof.__   On the line t_1 of the family $F_B(s_1)$ let D_1 and E_1 be
such that $<D_1BE_1>$ and on the line t_2 of $F_B(s_2)$ let D_2 and E_2
be such that $<D_2BE_2>$.   The rays $R[BD_i)$ have corresponding rays
$R[B'D_i^*)$,i=1,2 and the rays $R[BE_i)$ have corresponding rays
$R[B'E_i^*)$, i=1,2, and $\sphericalangle D_1BD_2$ corresponds with $\sphericalangle D_1^*B'D_2^*$.   Let T_i
denote a circle in $F_B(s_i)$ which does not pass through A,
i=1,2, and on T_i, let $arc_1(BP_i)$ be tangent to $R[BD_i)$, i=1,2.
By Cor. 6.2, $arc_1(BP_i)$ must maps to $arc_1(B'P_i')$ tangent to
$R[B'D_i^*)$, i=1,2.   Since the circle $T_i' = T_i\Phi_A(r)$ is tangent to
$L(B'D_i^*)$, $L(B'D_i^*)$ is the line of the family $F_{B'}(T_i')$, i=1,2.
Since $F_{B'}(S_i') = F_{B'}(T_i')$, it follows that $L(B'D_i^*)$ is the line
of $F_{B'}(s_i')$, i=1,2.   Therefore $\sphericalangle D_1^*B'D_2^*$, which corresponds with
$\sphericalangle D_1BD_2$, is an angle of intersection of s_1' and s_2'.   By Cor. 6.3,
the opposite angle of intersection at B', namely $\sphericalangle E_1^*B'E_2^*$,
corresponds with $\sphericalangle E_1BE_2$.   Clearly, the intersection angle
$\sphericalangle D_1^*B'E_2^*$ corresponds with $\sphericalangle D_1BE_2$, and $\sphericalangle D_2^*B'E_1^*$ corresponds with
$\sphericalangle D_2BE_1$.   □

__Convention.__   If an inversion maps two non-tangent curves s_1
and s_2 in a line-circle family F_B to curves s_1' and s_2'
respectively in the family $F_{B'}$, then an angle of intersection
of s_1 and s_2 at B and an angle of intersection of s_1' and s_2'
will be called corresponding angles of intersection if they
are corresponding angles in the inversion.

Corollary 8.1 If one intersection angle of s_1 and s_2 at B
is congruent to the corresponding intersection angle of s_1' and
s_2' at B', then each intersection angle at B is congruent to
the corresponding intersection angle at B'.

Corollary 8.2 If the intersection angles of s_1 and s_2 at B
are congruent to the corresponding intersection angles of s_1'
and s_2' at B', and if $s_3 \in F_B(s_1)$ and $s_4 \in F_B(s_2)$, then the intersection angles of s_3 and s_4 at B are congruent to the corresponding intersection angles of s_3' and s_4' at B'.

Because of Th. 8 and its corollaries, our initial objective
will be attained if we can show that corresponding angles in an
inversion are congruent. To do so we will make use of a familiar property of euclidean geometry which we can express in
terms of arc-ray tangency.

Theorem 9. If two circles c_1 and c_2 intersect at two points
P,Q and if arc_{11}(PQ) on c_1 is tangent to R(PX$_1$) and to R(QY$_1$),
and arc_{21}(PQ) on c_2 is tangent to R(PX$_2$) and to R(QY$_2$), then
$\not{*}X_1PX_2 \cong \not{*}Y_1QY_2$ and $\not{*}X_1PQ \cong \not{*}Y_1QP$.

Proof. Let u denote the line which is the perpendicular
bisector of S[PQ]. Then u is a diameter line of c_1 and c_2 and
so the reflection Γ_u maps c_1 onto itself and c_2 onto itself.
Because Γ_u leaves each side of L(PQ) invariant, each of the
arcs arc_{11}(PQ) and arc_{21}(PQ) maps onto itself. Thus Γ_u maps
R[PX$_1$) onto R[QY$_1$) and maps R[PX$_2$) onto R[OY$_2$). Because $\int_u$
preserves angle measure, $(\not{*}X_1PX_2)\Gamma_u = \not{*}Y_1QY_2$ implies $\not{*}X_1PX_2 \cong$
$\not{*}Y_1QY_2$. Also, Γ_u maps R[PQ) onto R[QP), and $(\not{*}X_1PQ)\Gamma_u = \not{*}Y_1QP$

implies $\star X_1 PQ \cong \star Y_1 QP.$ □

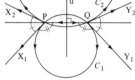

<u>Theorem 10.</u> If P and Q are distinct points which are inverse
with respect to a circle $C(A,r)$, and if s_1 and s_2 are two
curves which belong to both of the line-circle families F_p and
F_Q, then in the inversion $\Phi_A(r)$ the angles of intersection of
s_1 and s_2 at P are congruent to their corresponding angles of
intersection at Q.

<u>Proof.</u> Because P and its inversive image Q both belong to s_1
and to s_2, each of the curves s_1, s_2 is orthogonal to $C(A,r)$
and hence inverts onto itself. Since s_1 and s_2 cannot both be
lines, we may suppose that s_1 is a circle, and let $arc_{11}(PQ)$
denote an arc of P and Q on s_1. By Th. 2, there exist rays
$R(PX_1)$ and $R(QY_1)$ which are tangent to $arc_{11}(PQ)$ at P and at
Q respectively. Because the inversion maps $L(PQ)$ onto itself,
and leaves each side of $L(PQ)$ invariant, it maps $arc_{11}(PQ)$
onto itself. Thus, by Cor. 7.2, $arc_{11}(PQ)$ must be tangent at
Q to the open ray corresponding to $R(PX_1)$. Since the open ray
tangent to $arc_{11}(PQ)$ at Q is unique and is $R(PY_1)$, it follows
that $R(PY_1)$ and $R(PX_1)$ are corresponding.

If s_2 is also a circle, then $\text{arc}_{21}(PQ)$ on s has tangent
rays $R(PX_2)$ and $R(QY_2)$ and, by the same argument as before,
these must be corresponding rays. Thus $\star X_1 PX_2$ and $\star Y_1 QY_2$ are
corresponding intersection angles, and, by Th. 9, they are
congruent.

If s_2 is not a circle, then it is the line $L(PQ) = L(AP)$.
By definition, $\star X_1 PQ$ is an intersection angle of s_1 and s_2 at
P. By Cor. 7.1, $R(QP)$ is the ray corresponding to $R(PQ)$.
Thus $\star X_1 PQ$ and $\star Y_1 QP$ are corresponding intersection angles,
and, by Th. 9, they are congruent.

Because one angle of intersection of s_1 and s_2 at P is
congruent to its corresponding angle at Q, Cor. 8.1 implies
that each intersection angle at P is congruent to its corresponding angle.   □

In general, of course, we cannot expect that two curves
s_1 and s_2 in F_P will also belong to $F_{P'}$, where $P' = P\Phi_A(r)$.
However, if $s_3 \in F(s_1)$ and $s_4 \in F(s_2)$ then an intersection angle
of s_1 and s_2 at P is also an intersection angle of s_3 and s_4
at P. The next theorem implies that if P satisfies certain
restrictions, then s_3 and s_4 can be chosen so they do belong
to both F_P and $F_{P'}$.

Theorem 11. If point P is not A and is not on the circle
$C(A,r)$, then corresponding to each curve s in F_P there exists
a unique curve s^* in F which is orthogonal to $C(A,r)$.

Proof. By hypothesis, $P' = P\Phi_A(r)$ is not P. By Th. 10 and
Cor. 11, IV-2, the collection of all curves in F_P which are
orthogonal to $C(A,r)$ consists of $L(AP)$ and all circles passing
through P and P'. The locus of the centers of these circles

is the line u that is the perpendicular bisector of S[PP'].
Let t be the line in the family $F_P(S)$. Then $F_P(S)$ consists

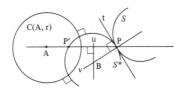

of t and all the circles which pass through P and have their
centers on the line v perpendicular to t at P. If u and v
are not parallel, they intersect at a point B, and the circle
$s* = C[B,d(B,P)]$ is the unique circle which passes through P
and P' and whose center is on v. Thus $s*$ is the unique curve
in $F_p(S)$ which is orthogonal to $C(A,r)$. If u and v are parallel, then there is no circle in $F_p(S)$ which is orthogonal to
$C(A,r)$. But $v \mid\mid u$ and $u \perp L(AP)$ implies that $v \perp L(AP)$.
Because $v \perp t$ at P and $v \perp L(AP)$ at P, it follows that $t =$
$L(AP)$. Thus $s* = L(AP)$ is the unique curve in $F_p(S)$ which is
orthogonal to $C(A,r)$.   □

   It is now easy to prove that if two corresponding angles
in an inversion have different vertices then they must be
congruent. Let $\sphericalangle XPY$ and $\sphericalangle X_1P'Y_1$ denote two such angles in an
inversion $\Phi_A(r)$, where $P' = P\Phi_A(r)$ is not P. Define $t_1 = L(PX)$
and $t_2 = L(PY)$. By Th. 11, there exist $s_1^* \in F_p(t_1)$ and $s_2^* \in F_p(t_2)$
such that s_1^* and s_2^* are orthogonal to $C(A,r)$. Since the inversion leaves s_1^* and s_2^* invariant, P' belongs to both s_1^* and s_2^*,
and so s_1^* and s_2^* belong to both F_p and $F_{p'}$. Because $s_1^* \in F_p(t_1)$
and $s*_2 \in F_p(t_2)$, $\sphericalangle XPY$ is an intersection angle at P of s_1^* and
s_2^*. By Th. 10, $\sphericalangle XPY$ is congruent to the corresponding intersection angle of s_1^* and s_2^* at P'. Because $t_1 \in F_p(s_1^*)$ and $t_2 \in F_p(s_2^*)$,

it follows from Cor. 8.2 that the intersection angles of t_1
and t_2 at P are congruent to the corresponding intersection
angles of $t_1\Phi_A(r)$ and $t_2\Phi_A(r)$ at P'. Therefore $\sphericalangle XPY \cong \sphericalangle X_1PY_1$.

To settle the remaining case in which two corresponding
angles in an inversion have the same vertex, we need a special
property from euclidean geometry. Two rays R(AX) and R(BY)
which are not collinear are defined to be like directed if
their lines are parallel and if the rays lie in the same side
of L(AB). It is a theorem of euclidean geometry that if the
arms of one angle are like directed respectively to the arms
of a second angle, then the angles are congruent.

Suppose now that P is on C(A,r) and that $\sphericalangle XPY$ and $\sphericalangle X_1PY_1$
are corresponding angles in the inversion $\Phi_A(r)$, with R(PX)
corresponding to R(PX$_1$) and R(PY) corresponding to R(PY$_1$).
Since both arms of $\sphericalangle XPY$ cannot be on L(AP), we may suppose
that R(PX) is not on L(AP). The line L(PX) = t_1 does not
pass through A and hence inverts to a circle C_1 through A.

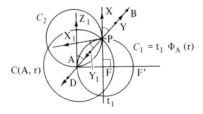

If F is the foot of A in t_1 and F' = $F\Phi_A(r)$, then S[AF'] is

a diameter of c_1. Let $arc_{11}(AP)$ denote the arc of A and P
on c_1 which lies in the X-side of L(AP) and let $R(AZ_1)$ be
the tangent to this arc at A. The line $L(AZ_1)$ is tangent
to c_1 at A, hence is perpendicular to L(AF), and is therefore
parallel to L(PX). Because $R(AZ_1)$ and R(PX) lie in the same
side of L(AP), and their lines are parallel, they are like
directed rays.

   Now consider the circle c_2 which passes through A and
is tangent to L(PX) at P. Let $arc_{21}(AP)$ denote the arc of A
and P on c_2 which lies in the X-side of L(AP). By Cor. 7.1,
$arc_{21}(AP)$ inverts to $R(PX_1)$ corresponding to R(PX). Thus c_2
inverts to $L(PX_1)$, and since c_2 and L(PX) are tangent at P,
their images are tangent at P. Therefore $L(PX_1)$ is tangent
at P to c_1. Because $\Phi_A(r)$ leaves the sides of L(AP) invariant,
$arc_{21}(AP)$ and $R(PX_1)$ both lie in the X-side of L(AP). Therefore $R(PX_1)$, which is tangent to one of the opposite arcs of
A and P on c_1, must be tangent to $arc_{11}(AP)$.

   If R(PY) is on L(AP), it must be either R(PA) or R(PB),
where <APB>. Suppose that R(PY) = R(PB). Then $\not{\star}XPY = \not{\star}XPB$.
Because R(AP) and R(PB) are like directed on L(AP), and $R(AZ_1)$
and R(PX) are parallel and like directed, it follows that
$\not{\star}Z_1AP^O = \not{\star}XPB^O = \not{\star}XPY^O$. Since $R(AZ_1)$ and $R(PX_1)$ are the tangent rays to $arc_{11}(AP)$ at A and P respectively, Th. 9 implies
that $\not{\star}Z_1AP^O = \not{\star}X_1PA^O$. But, by Cor. 7.1, R(PA) is the ray
$R(PY_1)$ corresponding to R(PY). Therefore $\not{\star}Z_1AP^O = \not{\star}X_1PY_1^O$,
and so $\not{\star}XPY^O = \not{\star}X_1PY_1^O$. If R(PY) = R(PA), then $\not{\star}XPY = \not{\star}XPA$.
If D on L(AP) is such that <PAD>, then R(PA) and R(AD) arc like
directed on L(AP), and so $\not{\star}XPY^O = \not{\star}Z_1AD^O = 180^O - \not{\star}Z_1AP^O$. By
Th. 9, $\not{\star}Z_1AP^O = \not{\star}X_1PA^O = 180^O - \not{\star}X_1PB^O$. But, by Cor. 7.1,
$R(PB) = R(PY_1)$. Therefore $\not{\star}Z_1AP^O = 180^O - \not{\star}X_1PY_1^O$.

From $\angle XPY^\circ = 180^\circ - \angle Z_1AP^\circ = 180^\circ - (180^\circ - \angle X_1PY_1{}^\circ)$, it follows that $\angle XPY^\circ = \angle X_1PY_1{}^\circ$. Thus the corresponding angles are congruent if $R(PY)$ is on $L(AP)$.

If $R(PY)$ as well as $R(PX)$ is not on $L(AP)$, then the previous arguments about figures associated with $R(PX)$ can be duplicated for $R(PY)$. That is, $L(PY)$ inverts to a circle c_1^*, and on c_1^* there exists $arc_{11}^*(AP)$ in the Y-side of $L(AP)$. The ray $R(AZ_2)$ which is tangent to $arc_{11}^*(AP)$ is like directed to $R(PY)$. The circle c_2^*, which passes through A and is tangent to $L(PY)$ at P, contains an arc $arc_{21}^*(AP)$ in the Y-side of $L(AP)$, and this arc inverts to the ray $R(PY_1)$ corresponding to $R(PY)$. The tangency of c_2^* and $L(PY)$ implies that $L(PY_1)$ is tangent to c_1^* and that $R(PY_1)$ is tangent to $arc_{11}^*(AP)$. Applying Th. 9 to the circles c_1 and c_1^*, which intersect at A and at P, it follows that $\angle Z_1AZ_2^\circ = \angle X_1PY_1^\circ$. But since $R(AZ_1)$ and $R(PX)$ are like directed, and $R(AZ_2)$ and $R(PY)$ are like directed, $\angle Z_1AZ_2^\circ = \angle XPY^\circ$, and therefore, $\angle X_1PY_1^\circ = \angle XPY^\circ$.

The foregoing arguments imply the angle measure invariance which we wanted to establish.

Theorem 12.

Corresponding angles in an inversion are congruent.

We turn now to a different aspect of inversions, namely how the mapping $X' = X\Phi_A(r)$ affects distances. First, it is a straightforward matter to calculate how $d(P,Q)$ and $d(P',Q')$ are related in terms of $d(A,P)$, $d(A,Q)$, and r.

<u>Theorem 13.</u>

If P' and Q' are the inverses of P and Q respectively with respect to the circle $C(A,r)$, then

$$d(P',Q') = \frac{r^2\, d(P,Q)}{d(A,P)d(A,Q)} \ .$$

<u>Proof</u>

If $P = Q$ then $P' = Q'$ which implies that $d(P',Q') = 0$, and this is also the value of $d(P',Q')$ given by the formula. If $P \neq Q$, and if A,P, and Q are non-collinear, then, by Th. 3, IV-2, $\triangle APQ \sim \triangle AQ'P'$, which implies that

$$d(P',Q')/d(Q,P) = d(A,P')/d(A,Q). \qquad (1)$$

Since P and P' are inverses with respect to $C(A,r)$, $d(A,P') = r^2/d(A,P)$, and this equality, together with (1), implies

$$d(P',Q') = \frac{r^2 d(P,Q)}{d(A,P)d(A,Q)} \ . \qquad (2)$$

If $P \neq Q$, and if A,P, and Q are collinear, we may suppose that $<APQ>$ and hence that $<AQ'P'>$.   From $<AQ'P'>$ it follows that

$$d(P',Q') = d(A,P') - d(A,Q'). \qquad (3)$$

Using $d(A,P') = r^2/d(A,P)$ and $d(A,Q') = r^2/d(A,Q)$, (3) can be expressed as

$$d(P',Q') = \frac{r^2}{d(A,P)} - \frac{r^2}{d(A,Q)} \ ,$$

or,

$$d(P',Q') = \frac{r^2[d(A,Q) - d(A,P)]}{d(A,P)d(A,Q)} \ . \qquad (4)$$

Since $<APQ>$ implies that $d(A,Q) - d(A,P) = d(P,Q)$, (4) is equivalent to

$$d(P',Q') = \frac{r^2 d(P'Q')}{d(A,P)d(A,Q)} \qquad (5) \qquad \square$$

In the formula of Th. 12, the denominator product is a clue to an inversion invariant related to four points. To see this, suppose that P,Q,R,S are points with inversive

images P',Q',R',S; respectively. The distance between two of

the four points, say P and Q, is changed by the inversion from

d(P,Q) to

$$d(P',Q') \ = \ \frac{r^2 d(P,Q)}{d(A,P)d(A,Q)} \ , \qquad (1)$$

and the distance between the other pair, R and S, is changed

to

$$d(R',S') \ = \ \frac{r^2 d(R,S)}{d(A,R)d(A,S)} \ . \qquad (2)$$

Thus if k is the product of the distances of the four points

from A, i.e. if k = d(A,P)d(A,Q)d(A,R)d(A,S), it follows from

(1) and (2) that

$$d(P',Q')d(R',S') \ = \ \frac{r^4}{k} \ d(P,Q)d(R,S). \qquad (3)$$

Thus the product of the two distances selected, namely

d(P,Q)d(R,S), is changed by the constant factor r^4/k. If

instead of choosing the pairs P,Q and R,S, different pairs

are selected, say P,R and Q,S, the same calculations show that

$$d(P',R')d(Q',S') \ = \ \frac{r^4}{k} \ d(P,R)d(Q,S). \qquad (4)$$

From (3) and (4), it follows that the ratio of the first dis-

tance product, d(P,Q)d(R,S), to the second distance product,

d(P,R)d(Q,S), is not changed by the inversion, since

$$\frac{d(P',Q')d(R',S')}{d(P',R')d(Q',S')} \ = \ \frac{d(P,Q)d(R,S)}{d(P,R)d(Q,S)} \ . \qquad (5)$$

The ratio of the two distance products on the right of (5) is

called a "cross ratio" of the points P,Q,R,S. This cross

ratio can be defined more precisely as follows.

Definition. (Cross ratio) The _cross_ _ratio_ of four points

_in_ _a_ _given_ _order_ is the product of the distances between the

first and third and the second and fourth divided by the product of the distances between the first and second and the third and fourth. Thus if P, Q, R, S are assigned the order (P,R,Q,S), their cross ratio in this order is

$$CR(P,R,Q,S) \;=\; \frac{d(P,Q)d(R,S)}{d(P,R)d(Q,S)} \;.$$

The equality (5), established prior to the definition, implies the following theorem.

Theorem 14.   In a circular inversion, the cross ratio of four points in a given order is the same as the cross ratio of their images in the corresponding order.

Exercises - Section 3

1. Prove Cor. 1.1 and Cor. 1.2.

2. Prove Th. 2.

3. Prove Th. 3.

4. Prove Th. 4.

5. Prove Th. 5.

6. Prove Th. 6.

7. Prove Cor. 7.3.

8. In the proof for Th. 11, a diagram is shown for the case in which u and v intersect. Make a diagram showing s and t for the case in which u is parallel to v.

9. In the inversion $X' = X\Phi_A(r)$, a circle $C(P,a)$, which does not pass through A, inverts to a circle $C(Q,b)$. Let u be any line through P and distinct from $v = L(AP)$. Because P is on both u and v, P' must be on both u' and v'. Use this property to show that $P' \neq Q$ and hence that the center of $C(P,a)$ does not invert to the center of the image circle.

10. Let P and P' denote inverses with respect to a circle $C(A,r)$, with P interior to $C(A,r)$. There exists a line $L(P'T)$ which is tangent to $C(A,r)$ at T. If $d(P',T) = b$, explain why the inversion in $C(P',b)$ must map P onto A.

11. A point P is given which is not on either of two given circles c_1 and c_2 which intersect at a point A. Explain how inversions can be used to find a circle which passes through P and is tangent to both c_1 and c_2.

12. Show that :
    (i) $CR(P,Q,R,S) = \dfrac{1}{CR(Q,P,R,S)} = \dfrac{1}{CR(P,Q,S,R)}$;

    (ii) $CR(P,Q,R,S) = CR(R,S,P,Q) = CR(Q,P,S,R)$.

Section 4. Linear Order and Motions in the Model

In Section 1, the interior of a circle $C(0,1)$ in E^2 was named an "h-space". Certain subsets of this space were named "lines" and a rule was given for assigning a number, called an "h-distance", to pairs of points in the h-space. To show that this nominal system is actually a representation of hyperbolic geometry, it must be shown that for each axiom of hyperbolic geometry about lines, distance, angles and triangles, there is a counterpart h-theorem about h-lines, h-distance, h-angles and h-triangles. We will call such a counterpart an "axiom-h-theorem" to indicate that while it is a euclidean theorem about objects in E^2, which have been given h-names, it has the property that if the h-space qualifications are deleted from the theorem it becomes the statement of an axiom in hyperbolic geometry. In this section we will concentrate on the axiom h-theorems relating to lines and to h-distance.

The first axiom of absolute geometry, hence of hyperbolic geometry, given in II-1, was the statement "There exist non-empty subsets of the plane, called "lines", with the property that each two points belong to exactly one line." The following is the corresponding axiom-h-theorem.

Theorem 1.   There exist non-empty subsets of the h-space, called "h-lines", with the property that each two h-points belong to exactly one h-line.

Proof

By their definition, h-lines exist and are non-empty sets. To show that each two h-points A, B belong to exactly one h-line, suppose first that A and B are collinear with O. Then L(AB) carries a central h-line, say t, to which A and B belong. There

is no other central h-line to which A and B belong, since two
such lines intersect only at O. If a non-central h-line passed
through A and B, then, by Cor. 11, IV-2, A and B would be inverse
with respect to c^*, contradicting the fact that both are
interior to c^*.   Thus t is the only h-line to which A and B
belong.

Suppose next that A and B are not collinear with O. Then
no diameter line of c^* contains A and B, and so no central h-
line passes through A and B. The circumcircle c of $\triangle AA'B$
passes through A and A' and hence, by Th. 10, IV-2, is orthogonal to c^*. Thus c is the carrier of an h-line t through A

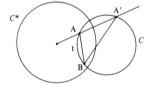

and B. If s denotes an h-line through A and B, s cannot be
central and so its carrier c' is a circle orthogonal to c^*.
By Cor. 11, IV-2, L(OA) intersects c' at A and at the point
which is the inverse of A with respect to c^*, namely A'. Thus
c' is the circumcircle of $\triangle AA'B$, hence $c' = c$, and so,
$s = t$. Thus t is the unique h-line through A and B.   □

Corollary 1.

Two h-lines intersect at at most one point.

The following theorem will be extremely useful.

Theorem 2.

Each two points of $c*$ belong to the carrier of exactly one h-line.   (Ex.)

Conventions

If A and B are two h-points, the notation $L_h(AB)$ will denote the h-line of A and B. We will also, from now on, use the star notation exclusively for points of $c*$. Thus "point A*" will automatically imply that A* $\in$ $c*$.   If P* and Q* are distinct points of $c*$, the notation $L_h(P*Q*)$ will denote the h-line whose carrier intersects C* at P* and Q*. Thus P* and Q* do not belong to $L_h(P*Q*)$ and if A and B are two points that do belong to $L_h(P*Q*)$ they are h-points and $L_h(AB) = L_h(P*Q*)$.

Turning now to h-distance, we will adopt the base e for the logarithm defining $h(A,B)$.* Thus, if A and B are distinct h-points, they determine a unique h-line which intersects $c*$ at a pair of points P* and Q*, and the h-distance of A and B is the number

$$h(A,B) = \left| Ln \left[\frac{d(P*,B)\ d(Q*,A)}{d(P*,A)d(Q*,B)} \right] \right| = \left| Ln \left[CR(P*,A,B,Q*) \right] \right|. \quad (1)$$

From the properties of cross ratio, (cf. exercises, IV-3),

$$h(B,A) = \left| Ln \left[CR(P*,B,A,Q*) \right] \right| = \left| Ln \left[CR(P*,A,B,Q*)^{-1} \right] \right|$$
$$= \left| -Ln \left[CR(P*,A,B,Q*) \right] \right| = \left| Ln \left[CR(P*,A,B,Q*) \right] \right|$$
$$= h(A,B). \quad (2)$$

Thus h-distance has the necessary symmetry property.

The absolute values used in the definition (1) ensure that h-distance is a non-negative number. But since cross ratios, by definition, are positive, not all the absolute values in (2)

---

*   The notation "Ln x" is commonly used for "$\log_e x$".

can be necessary. The cross ratios $CR(P*,A,B,Q*)$ and
$CR(P*,B,A,Q*)$ are reciprocals, thus one of them is equal to or
less than 1 and the other is equal to or greater than 1, so
the logarithm of one of them must be non-negative. If $L_h(A,B)$
is a central h-line, then A and B belong to $S(P*Q*)$, and so
$A,B,P*,Q*$ are successive on $L(P*Q*)$ in one of the orders $(P*,A,B,Q*)$,
$(P*,B,A,Q*)$, and we may suppose it is the order $(P*,A,B,Q*)$.
From $<P*AB>$ and $<ABQ*>$ it follows that $d(P*,B)/d(P*,A) > 1$
and $d(Q*,A)/d(Q*,B) > 1$. Therefore $CR(P*,A,B,Q*) =$
$CR(Q*,B,A,P*) > 1$, and so

$$h(A,B) = Ln [CR(P*,A,B,Q*)] = Ln [CR(Q*,B,A,P*)] > 0. \quad (3)$$

To establish the formula (3) for non-central h-lines, we
need a few more arc properties.

Definition

(Betweeness and successive order on an arc)   Point B is
__between__ points A and C __on__ __an__ __arc__ s if arc(AC) is contained in
s and B belongs to arc(AC). An ordering $(P_1,P_2,...,P_n)$ of n
points on an arc s is a __successive__ order __on__ __the__ __arc__ if P_i is
between P_{i-1} and P_{i+1} on the arc s, $i = 2,3,...,n-1$.

The following agreements about arcs related to h-lines will
be convenient.

Conventions

The use of the capital A in "Arc" will be used to denote
an arc which belongs to the circular carrier of an h-line and
which contains no points exterior to $c*$. Thus if $L_h(AB)$ is
non-central, with carrier c, Arc(AB) denotes the arc of A
and B on c which is contained in H. If $L_h(AB) = L_h(P*Q*)$,

then $L_h(AB) = Arc(P*Q*)$. If E is the center of the carrier c ,
$Arc(P*Q*)$ is in the non-E-side of $L(P*Q*)$ and is interior
to $\angle P*EQ*$. Thus all "capital arcs" are minor arcs.

If arc[PQ] is a minor arc on a circle with center E
and if point A is between P and Q on the arc then clearly
$R(EA) \subset In(\angle PEQ)$.   These relations, together with Th. 11,
II-3, imply the following theorem.

<u>Theorem 3.</u>

If A is between P and Q on a minor arc of a circle then
$d(P,Q) > d(P,A)$ and $d(P,Q) > d(Q,A)$.   (Ex.)

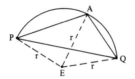

Returning to the model, let A and B denote two points of
a non-central h-line, with $L_h(AB) = L_h(P*Q*)$ and with P* and Q*
labeled so that $(P*,A,B,Q*)$ is an order in which the points are
successive on Arc [P*Q*] .   Then, by Th. 3,

$$\frac{d(P*,B)}{d(P*,A)} > 1 \quad \text{and} \quad \frac{d(Q*,A)}{d(Q*,B)} > 1.$$

Therefore,

$$CR(P*,A,B,Q*) = CR(Q*,B,A,P*) = \frac{d(P*,B)}{d(P*,A)}\frac{d(Q*,A)}{d(Q*,B)} > 1,$$

and so

$$h(A,B) = \text{Ln}[\,CR(P^*,A,B,Q^*)\,] = \text{Ln}[\,CR(Q^*,B,A,P^*)\,] > 0. \quad (4)$$

From equations (3) and (4) it follows that if two h-points and the corresponding boundary points on their h-line are assigned an order in which the points are successive on the carrier of the h-line, then the cross ratio of the points in this order is greater than 1 and the logarithm of the cross ratio of the points in this order is the distance between the h-points. In particular, (3) and (4) imply that if $A \neq B$, then $h(A,B) > 0$.

As yet, $h(A,B)$ has no meaning if $A = B$, so we introduce the following definition.

Definition

(Zero h-distance)   If A is an h-point, the distance of A from A, denoted by $h(A,A)$, is zero.

The definition just introduced, together with (3) and (4) imply the following axiom-h-theorem (cf. Axiom 2, II-1).

Theorem 4.

Corresponding to h-points A,B there exists a unique non-negative number $h(A,B) = h(B,A)$ which is the h-distance between A and B and which is zero if and only if $A = B$.

Betweeness of points in the sense of h-distance is defined in exactly the same way that betweeness in A^2 was defined in terms of the d-distance.

Definition

(h-betweeness)   Point B is between points A and C if

A,B,C are three points of an h-line and

$$h(A,B) + h(B,C) = h(A,C).$$

That B is between A and C, in an h-sense, will be denoted by
h- <ABC> or by h- <CBA> .

The next theorem shows that h-lines are 'straight' in
the sense that h-distances are additive on the lines of H .

Theorem 5.

If A,B,C are three points of an h-line, exactly one of them
is between the other two in an h-sense.

Proof

Let $t = L_h(P*Q*)$ be the h-line of A,B,C. From euclidean
geometry it follows that exactly one of the points A,B,C is
between the other two on S[P*Q*] , if t is a central line,
and exactly one is between the other two on Arc [P*Q*] if t
is a non-central line. Without loss of generality, we may
suppose that B is between A and C in the sense of the carrier.
Then one of the orders (P*,A,B,C,Q*) and (Q*,A,B,C,P*) is a
successive order (either on S[P*Q*] or Arc [P*Q*]). Since
the labels P* and Q* can be interchanged, we may suppose that
(P*,A,B,C,Q*) is a successive order of the points on the
carrier. Then (P*,A,B,Q*), (P*,B,C,Q*), and (P*A,C,Q*) are also
successive orders, and so

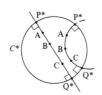

$h(A,B) = Ln [CR(P*,A,B,Q*)], h(B,C) = Ln [CR(P*,B,C,Q*)]$, and
$h(A,C) = Ln [CR(P*,A,C,Q*]$. Therefore,

$$h(A,B) + h(B,C) = Ln [CR(P*,A,B,Q*)] + Ln [CR(P*,B,C,Q*)]$$

$$= Ln [CR(P*,A,B,Q*) \cdot CR [(P*,B,C,Q*)]$$

$$= Ln [\frac{d(P*,B)d(Q*,A)}{d(P*,A)d(Q*,B)} \frac{d(P*,C)d(Q*,B)}{d(P*,B)d(Q*,C)}]$$

$$= Ln [\frac{d(P*,C)d(Q*,A)}{d(P*,A)d(Q*,C)}] = Ln [CR(P*,A,C,Q*)]$$

$$= h(A,C),$$

and so h-<ABC> .

Since $h(A,B)$, $h(B,C)$ and $h(A,C)$ are positive numbers,
h-<ABC>   implies that $h(A,C) > h(A,B)$ and that $h(A,C) >$
$h(B,C)$. Thus h-<BCA> is not possible, since it contradicts
$h(A,C) > h(A,B)$, and h-<CAB> is not possible, since it contradicts $h(A,C) > h(B,C)$. Therefore B is the only one of the
three points which is between the other two.   □

Corollary 5.

One h-point is between two others, in an h-sense, if and
only if it is also between the two others in the sense of the
carrier of their h-line.

The formula for calculating the h-distance of two points
takes a relatively simple form when one of the points is 0, the
euclidean center of H . Because of this, it is fairly easy to
establish a coordinate system for a central h-line which has the
properties of the ruler axiom. One can then transfer such a
coordinate system to a non-central h-line by means of a mapping
of H onto itself which preserves h-distance, i.e. by an
"h-motion". But to employ this strategy, we need some facts
about h-motions.

It is not difficult to see that any motion Γ of E^2 which leaves c^* invariant must map H onto itself in such a way that h-distances are preserved. First, since O is equidistant from all points of c^*, its image $O' = O\Gamma$ must be equidistant from all points of $c^*\Gamma = c^*$, hence $O' = O$. If $X \in H$, $d(O,X) < 1$ implies that $d(O,X\Gamma) < 1$, thus H maps into H. But if $Y \in H$ there is some point $Z \in E^2$ such that $Z\Gamma = Y$, and $d(O,Z) = d(O,Y) < 1$ implies that $Z \in H$. Since each point of H is the Γ-image of a point in H, it follows that H maps onto H. Since O is a fixed point and c^* is a fixed set and $H\Gamma = H$, a central h-line $t = L_h^{'}(P^*Q^*)$ must map to a central h-line t' whose carrier is the euclidean line through $P^*\Gamma$ and $Q^*\Gamma$. Because Γ preserves angle measure, if $t = L_h(P^*Q^*)$ is non-central its circular carrier c, which is orthogonal to c^* at P^* and Q^* must map to a circle c' which is orthogonal to C^* at $P^*\Gamma$ and $Q^*\Gamma$. Since h-points map onto h-points, it follows that t maps to the non-central h-line carried by c'. Finally, if A and B are points of t, whether t is central or non-central, the fact that Γ preserves euclidean distance implies that $CR(P^*,A, B,Q^*)= CR(P^*\Gamma, A\Gamma, B\Gamma, Q\Gamma,)$. Therefore,

$h(A,B) = |Ln[\,CR(P^*,A,B,Q^*)]\,| = |Ln[\,CR(P^*\Gamma, A\Gamma\;\; B\Gamma, Q^*\Gamma)]\,| = h(A\Gamma, B\Gamma)$. Thus Γ preserves h-distance.

The fact that the mapping of H onto itself by Γ is only part of a more extensive mapping is commonly expressed by saying that " Γ induces a mapping of H onto itself". We use this language in summarizing the previous conclusions.

Theorem 6

A euclidean motion of E^2 which leaves c^* invariant induces a mapping of H onto itself which maps h-lines onto h-lines and preserves the h-distance between points of *H* .

Corollary 6.1

A euclidean motion of E^2 which leaves c^* invariant preserves the h-betweeness of points in *H* .   (Ex.)

Corollary 6.2

The mappings of *H* onto itself induced by the reflection of E^2 in point O and by the reflection of E^2 in a diameter line of c^* are h-motions of *H* .

Another class of h-motions is obtained from circular inversions.

Theorem 7.

The inversion of E^2 in a circle $C(E,r)$ which is orthogonal to c^* induces a mapping of *H* onto itself in which h-lines map onto h-lines and h-distance is preserved.

Proof

Let L(EP*) and L(EQ*) be the lines through E which are tangent to C* at P* and Q* respectively. Then P* and Q* are fixed points of $\Phi_E(r)$ as are all points of the h-line $s = L_h(P^*Q^*) =$ Arc(P*Q*). If Z is an arbitrary point of *H*, it follows from

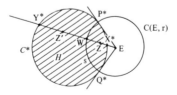

$H \subset \text{In}(\text{\textonehalf}P^*EQ^*)$ that R(EZ) intersects c^* at a pair of
points X*,Y* and, by Cor. 11, IV-2, they are inverses with res-
pect to C(E,r). We may suppose that $<$EX*Y*$>$. The point
W at which R(EZ) intersects s is a fixed point. Denoting images
by primes, if Z $\in$ S(X*W) then d(E,X*) $<$ d(E,Z) $<$ d(E,W)
implies that d[E,(X*)'] = d(E,Y*) $>$ d(E,Z') $>$ d(E,W') = d(E,W),
and so Z' $\in$ S(WY*). Clearly, $\Phi_E(r)$ maps S(X*Y*) onto
itself by leaving W fixed and interchanging S(WX*) and S(WY*).
Thus H maps onto itself, s is pointwise invariant, and the two
open h-half planes of s are interchanged.

    Now consider an arbitrary h-line t = L_h(U*V*). The inver-
sion maps c^* onto itself and therefore maps U* to a point U_1^*
on C* and maps V* to a point V_1^* on c^*. The set s, which is
the carrier of t, is either a line or a circle to which U* and
V* belong, so its inverse s' is either a line or a circle to
which U_1^* and V_1^* belong. Since s is orthogonal to c^* at U* and
V*, and since c^* is invariant, it follows from Th. 12, IV-3,
that s' is orthogonal to c^* at U_1^* and V_1^*. Therefore s' is
either a line or a circle which is orthogonal to c^*, and hence
is the carrier of an h-line. Because h-points invert to h-
points, it follows that t inverts to the h-line carried by s'.

    If A and B are two points of t, with inversive images A'
and B' respectively on t' = t $\Phi_A(r)$, then, by definition,

$$h(A,B) = |\text{Ln} [CR(U^*,A,B,V^*)] |$$

and   (1)

$$h(A',B') = |\text{Ln} [CR(U_1^*, A',B',V_1^*)]| \quad .$$

But, by Th. 14, IV-3, the two cross ratios in (1) are equal,
and therefore h(A,B) = h(A',B'). Thus the inversion preserves
h-distances. $\square$

Corollary 7.

The inversion preserves h-betweeness of points in H.   (Ex.)

We return now to the problem of establishing an h-equivalent
to the ruler axiom. First, consider an arbitrary central h-line
$t = L_h(P^*Q^*)$. Corresponding to each point X on t, let an h-
coordinate of X be the real number x defined by:

$$x = h(0,X) \text{ if } X \in S(OP^*);$$
$$x = 0 \text{ if } X = 0; \tag{1}$$
$$x = -h(0,X) \text{ if } X \in S(OQ^*).$$

We want to show that the correspondence X < - > x, X $\in$ t, is
a one-to-one correspondence between the points of t and the set
of all real numbers and that if Y, Z on t have h-coordinates y,z
respectively, then $h(Y,Z) = |y - z|$.

If $X \in S(OP^*)$, then $X \neq 0$ implies that $x = h(0,X) > 0$. Also,
if Y and Z are two points of $S(OP^*)$, then $d(0,Y) \neq d(0,Z)$ implies
that either < OYZ > or <OZY>. Since <OYZ> $\rightarrow$ h-<OYZ> $\rightarrow$ h(0,Y) < h(0,Z)$\rightarrow$
y < z, and <OZY> $\rightarrow$ h-<OZY> $\rightarrow$ h(0,Z) < h(0,Y) $\rightarrow$ z < y, it follows that
$Y \neq Z$ implies $y \neq z$. Thus each point of $S(OP^*)$ has a unique,
positive h-coordinate.

Corresponding to $X \in S(OP^*)$, let $\bar{x} = d(0,X)$. Since
$(Q^*,0,X,P^*)$ is a successive order of the points on $L(Q^*P^*)$,

$$x = h(0,X) = Ln[CR(Q^*,0,X,P^*)]$$
$$= Ln \left[\frac{d(Q^*,X)d(P^*,0)}{d(Q^*,0)d(P^*,X)}\right]. \tag{2}$$

Since $C^* = C(0,1)$, $d(Q^*,0) = d(P^*,0) = 1$. Also, < Q*OX> implies
that $d(Q^*,X) = d(Q^*,0) + d(0,X) = 1 + \bar{x}$, and <OXP*> implies
that $d(P^*,X) = d(P^*,0) - d(0,X) = 1 - \bar{x}$. Thus (2) can be put
in the form

$$x = h(0,X) = Ln(\frac{1 + \bar{x}}{1 - x}). \tag{3}$$

Now, let a be an arbitrary positive number. Then the number $\bar{b}$ defined by $\bar{b} = (e^a-1)/(e^a+1)$ is clearly a positive number which is less than 1. Therefore there exists a point B on S(OP*) such d(0,B) = $\bar{b}$. Because $B \in S(OP*)$, B has a positive h-coordinate b, and, by (3),

$$b = h(0,B) = Ln (\frac{1 + \bar{b}}{1 - \bar{b}}). \tag{4}$$

Since $\bar{b}$ was defined by $\bar{b} = \frac{e^a - 1}{e^a + 1}$, then $\bar{b}(e^a + 1) = e^a - 1$,

so $\bar{b} + 1 = e^a(1 - \bar{b})$, hence

$$\frac{1 + \bar{b}}{1 - \bar{b}} = e^a. \tag{5}$$

Together, (4) and (5) imply that

$$b = Ln(e^a) = a, \tag{6}$$

and hence that point B on S(OP*) has the h-coordinate a. Since each point of S(OP*) has a unique, positive coordinate, and since each positive number is the h-coordinate of a point on S(OP*), it follows that the correspondence X <-> x, $X \in S(OP*)$ is a one-to-one coorespondence between S(OP*) and the set of all positive real numbers.

If $X \in S(OQ*)$, $X \neq 0$ implies $h(0,X) > 0$, hence x = -h(0,X) < 0. If Y and Z are distinct points of S(OQ*) then, as argued before, $h(0,Y) \neq h(0,Z)$, hence $-h(0,Y) \neq -h(0,Z)$, so $y \neq z$. Thus each point of S(OQ*) has a unique, negative coordinate. If a is an arbitrary negative number then, by the same argument as before, there exists a point $C \in S(OQ*)$ such that h(0,C) = -a, and so the h-coordinate of C is c = -h(0,C) = a. Thus the correspondence X <-> x, $X \in S(0,Q*)$ is one-to-one between S(OQ*) and the set of all negative numbers.

Since 0 has coordinate zero, it follows from the last two paragraphs that each point X of t has a unique coordinate x defined by (1) and that each real number x is the coordinate of exactly one point X on t.

To show that $Y, Z \in t$ implies $h(Y,Z) = |y-z|$ is a tedious but straightforward matter of dealing with various cases.

Case 1. $Y = Z$. Then $y = z$. Since $Y = Z$, $h(Y,Z) = 0$, and since $y = z$, $|y-z| = 0$. Thus, $h(Y,Z) = |y-z|$.

Case 2. $Y \neq Z$ and $0 \in \{Y, Z\}$. We may suppose that $Y = 0$ and $Z \neq 0$. Then $h(Y,Z) = h(0,Z) = z$, if $Z \in S(OP*)$, and $h(Y,Z) = h(0,Z) = -z$, if $Z \in S(OQ*)$. Therefore $h(Y,Z) = |h(Y,Z)| = |z| = |z - y| = |y-z|$, (since $y = 0$).

Case 3. $Y \neq Z$, and $0 \notin \{Y, Z\}$. Now, by Cor. 5, there are three subcases, namely h-<YZO>, h-<ZOY>, and h-<OYZ>.

Subcase 3.1 h-<YZO>. Now, by definition,
$$h(Y,Z) = |h(Y,Z)| = |h(Y,0) - h(Z,0)| . \quad (7)$$
Since h-<YZO> implies <YZO>, either $Y, Z \in S(OP*)$ or $Y, Z \in S(OQ*)$. If $Y, Z \in S(OP*)$, then $y = h(Y,0)$, $z = h(Z,0)$ and (7) becomes $h(Y,Z) = |y-z|$. If $Y, Z \in S(OQ*)$, $h(Y,0) = -y$, $h(Z,0) = -z$, and (7) becomes $h(Y,Z) = |(-y)-(-z)| = |z-y| = |y-z|$.

Subcase 3.2 h-<ZOY>.   (Ex.)

Subcase 3.3 h-<OYZ>.   (Ex.)

Having shown that the correspondence X <-> x, defined in (1), has the properties of the ruler axiom when t is a central

h-line, suppose that t is a non-central h-line, say $t = L_h(P*U*)$.
Let Q* be the point of c* antipodal to P* and let line
s be the line tangent to c* at P*. The secant to c*, line
L(Q*U*), is not parallel to s and intersects it at a point E.
If $d(E,P*) = r$, the circle C(E,r) is orthogonal to c*, and
U*,Q* are inverses with respect to C(E,r). Since the inver-
sion $\Phi_E(r)$ maps U* to Q*, and leaves P* on C(E,r) fixed, it
maps the h-line t to the central h-line $t' = L_h(P*Q*)$. Let

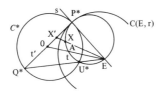

$A = 0 \Phi_E(r)$, so $0 = A'$, and let each point $X' \in t'$ be assigned
a coordinate x by the rules in (1) as before. That is, x =
h(0,X') if $X' \in S(OP*)$, x = -h(0,X') if $X' \in S(OQ*)$, and x = 0
if $X' = 0$. Finally, define the coordinate of point X on t to
be the coordinate x of X' on t'. Since X' <-> x is a one-to-
one correspondence of t' and the real numbers, and X' <-> x is
a one-to-one correspondence of t' with t, then X <-> x is a
one-to-one correspondence of t and the real numbers. By Th.7,
the inversion $\Phi_E(r)$ preserves h-distances, hence if Y and Z
are points of t, then $h(Y,Z) = h(Y',Z')$. We have already
proved that $h(Y',Z') = |y-z|$, and therefore $h(Y,Z) = |y-z|$.
This completes the proof for the following axiom-h-theorem
corresponding to the ruler axiom, (c.f. Axiom 3, II-1).

Theorem 8. If t is an h-line and R is the set of all real
numbers, there exists a one-to-one correspondence, denoted
by X < - > x, between the h-points X on t and the numbers x
in R such that the h-distance between points A, B on t is the
absolute value of the difference of the numbers a, b in R
which correspond to A and B respectively.

With Th. 8, we now have h-equivalents for the first three
axioms of absolute geometry on which theorems 1 to 12 in II-1
were based. Using h-betweeness, we can paraphrase the definitions of segments and rays to define h-segments and h-rays.
Having done so, the h-statements corresponding to theorems 1
to 12 in II-1 need not be proved since they are automatically
theorems.

We will denote the open and closed h-segments of h-points
A and B by $S_h(AB)$ and $S_h[AB]$ and the (ordered) half-open h-
segments by $S_h[AB)$ and $S_h(AB]$. The open h-rays from the h-
point A through the h-point B will be denoted by $R_h(AB)$ and
the corresponding closed ray by $R_h[AB)$. The h-definitions
for these sets as well as "opposite rays" and "like and opposite directed rays" is left to the reader. As one example of
a "free" theorem in the model, the following is implied by
Th. 6, II-1.

Theorem 9. If k is a positive number, there is exactly one
point C on $R_h(AB)$ whose h-distance from A is k, and h-<ABC>,
C = B, or h-<ABC> according as $h(A,C) < (A,B)$, $h(A,C) = h(A,B)$,
or $h(A,C) > h(A,B)$.

It is not surprising that the euclidean space in which
the model is embedded provides convenient ways of expressing
h-concepts which are not available in H^2. An example is the

notation $L_h(P*Q*)$ for an h-line. A similar convenience for
h-rays can be obtained from the following theorems.

<u>Theorem 10</u>.   Corresponding to $A \in H$ and $P* \in C*$, there is
exactly one h-line whose carrier passes through A and P*. (Ex.)

<u>Convention</u>   Corresponding to an h-point A and a boundary
point **P***, the unique h-line whose carrier passes through A
and P* will be denoted by $L_h(AP*)$ or $L_h(P*A)$. According as
$L_h(P*A)$ is central or non-central, $S(AP*)$ or $Arc(AP*)$ is an
open h-ray and will be denoted by $R_h(AP*)$. The corresponding
closed h-ray with origin A is $R_h[AP*)$. We will call $R_h(AP*)$
and $R_h[AP*)$ the open and closed h-rays at A in the direction
P* .

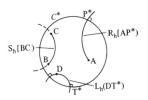

For the h-equivalents of Th. 13 and Th. 14 of II-1, we
need the following natural concept.

<u>Definition</u>.   (h-convexity) A set s contained in H is <u>h-convex</u>
if $A \in s$, $B \in s$ and h-$<AXB>$ imply that $X \in s$.

The h-equivalent of Th. 13 is the property that the intersection set of any number of h-convex sets is itself an h-convex

set. From Th. 14, II-1, it follows that the space H, the
empty set, singleton sets in H, and lines, rays, and segments
in H are h-convex sets.

<u>Exercises - Section 4.</u>

1.   Prove Th. 2.

2.   Prove Th. 3.

3.   Prove Cor. 6.1

4.   Prove Cor. 7.

5.   Prove subcases 3.2 and 3.3 in the argument preceding
     Th.8.

6.   Give definitions for the segements $S_h(AB)$, $S_h[AB]$, $S_h[AB)$,
     and for the rays $R_h(AB)$, $R_h[AB)$.

7.   Define "opposite h-rays", "like directed h-rays, and
     "opposite directed h-rays".

8.   Prove Th. 10.

9.   In the proof for Th. 8 a method was given for finding
     an h-motion that mapped the non-central h-line $t = L_h(P*U*)$
     onto the parallel, central h-line $L_h(P*Q*)$. If $u = L_h(P*V*)$
     is a non central h-line parallel to t, define an h-motion
     that maps t onto u.

10. Define the "midpoint" M of $S_h[AB]$. How do you know that
     M exists? If $L_h(AB)$ is non-central, give a geometric
     method for finding M.

Section 5. Half-Planes, Angles and Angle Measure in the Model

In this section, we will establish h-equivalents for
the plane separation axiom, the protractor axiom, and the
angle addition axiom. We begin with a natural definition.

Definition. (Separation of sets by an h-line) Sets R and
S in H are separated by the h-line t if R is contained in one
side of t and s is contained in the opposite side of t.
Points A and B are separated by t if they belong to opposite
sides of t.

Conventions: The opposite sides of line t in H will be denoted by $H_1(t)$ and $H_2(t)$, and $H_1[t]$ and $H_2[t]$ will denote the
corresponding closed half-planes.

The following property will be useful.

Theorem 1. A central h-line t separates h-sets R and S if
and only if the line carrying t separates R and S. (Ex.)

If t is an h-line, some of the properties in the plane
separation axiom (Axiom 4, II-2) have h-equivalents satisfied
by t and its sides as a matter of definition. Thus $H_1(t)$ and
$H_2(t)$ are non-empty sets and, by definition, H is the union
of t, $H_1(t)$, and $H_2(t)$. Also, the definitions of t, $H_1(t)$,
and $H_2(t)$ imply that no two of these sets intersect. However,
it remains to be proved that: $H_1(t)$ and $H_2(t)$ are h-convex;
that $X \in H_1(t)$ and $Y \in H_2(t)$ imply $S_h(XY) \cap t \neq \emptyset$; and that
$H_1(t)$ and $H_2(t)$ are the only two sets which satisfy all these
conditions.

In establishing the properties just described, and in
many other contexts, it is simpler to deal with a line which

is central rather than non-central. It is for this reason that
the circular inversion in the following theorem plays a key
role throughout this section.

Theorem 2. If an h-point A is not 0, and if B is the inverse
of A with respect to $c*$, the inversion in the circle $C(B,r)$
which is orthogonal to $c*$ induces a motion of H in which A
maps to 0. The h-line t whose carrier c_t is orthogonal to
L(OA) at A maps to a central h-line t_1 whose carrier L is
perpendicular to L(OA) at 0. The side $H_1(t)$ interior to c_t
maps onto the side $H_1(t_1)$ which is contained in the non-B-side
of L and $H_2(t)$ maps onto $H_2(t_1)$.

Proof. Since B is exterior to $c*$, there exists a line L(BT*)
tangent to $c*$ at T*, and the circle $C(B,r)$, $r = d(B,T*)$, is
orthogonal to $c*$. Thus the inversion $\Phi_B(r)$ induces a motion
of H . By Th. 2, IV-2, 0 and A are inverses with respect to
$C(B,r)$ and hence are interchanged by $\Phi_B(r)$. The circle with
diameter S[AB] is orthogonal to $c*$, since it passes through
A and B, and is therefore a circle c_t which carries an h-line
t. Let P* and Q* be the intersection points of $c*$ and c_t,
and let $P*\Phi_B(r) = P_1^*$ and $Q*\Phi_B(r) = Q_1^*$. Because B $\in c_t$, the
circle c_t inverts to a line which must be $L(P_1^*Q_1^*)$. Thus t_1
$= t_{\Phi_B}(r) = L_h(P_1^*Q_1^*)$ is a central h-line. Since L(OA) is a
diameter line of c_t it is orthogonal to c_t at A, hence its
image $L(P_1^*Q_1^*)$ is perpendicular to L(OA) at 0 (since L(OA) is
invariant).

Let $arc_1(P*Q*)$ be the arc on $c*$ which is interior to c_t .
If X $\in H_1(t)$, then R(BX) $\subset$ In($\not< P*BQ*$) which implies that R(BX)
intersects $arc_1(P*Q*)$ at a point Y* and intersects t at a

point Z.

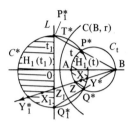

Let Y_1^*, X_1 and Z_1 denote the images of Y*, X, and Z respectively. Clearly, $c_t \subset \text{In}[C(B,r)]$, hence $d(B,Y^*) < d(B,Y_1^*)$.
Because $X \in H$, X belongs to $S(Y^*,Y_1^*)$, and so $d(B,Y^*) < d(B,X)$.
Since $X \in H_1(t)$, X belongs to $S(BZ)$, hence $d(B,X) < d(B,Z)$.
These inequalities imply that $d(B,Y^*) < d(B,X) < d(B,Z)$, and therefore, in turn, that $d(B,Y_1^*) > d(B,X_1) > d(B,Z_1)$. Since $Z_1 \in t_1$,
it follows that Y_1^* belongs to the open semicircle $\text{arc}_1(P^*Q^*)$
on c^* in the non-B-side of $L(P_1^*Q_1^*)$ and that X_1 belongs to the
side $H_1(t_1)$ which has this semicircle as its c^*-boundary.
So $H_1(t)$ maps into $H_1(t_1)$. But the argument is reversible.
That is, if $X \in H_1(t_1)$, then $R(BX)$ intersects t_1 at a point
Z and $\text{arc}_1(P_1^*Q_1^*)$ at a point Y*. From $d(B,Y^*) > d(B,X) > d(B,Z)$ it
follows that $d(B,Y_1^*) < d(B,X_1) < d(B,Z_1)$, and hence that X_1
$\in H_1(t)$. Thus $H_1(t)$ maps onto $H_1(t_1)$ and therefore $H_2(t)$
maps onto $H_2(t_1)$.   □

Corollary 2. If t is a non-central h-line with carrier c_t,
the common diameter line of c^* and c_t intersects c_t at the

points A,B of Th. 2 and t is the h-line which the inversion
$\Phi_B(r)$ of Th. 2 maps to a central h-line whose carrier is perpendicular to L(OA). (Ex.)

Theorem 3. If the h-point A is not O, then the circular
inversion of Th. 2 is the only inversion in a circle orthogonal to c^* which maps A to O. (Ex.)

Convention. The inversion in Th. 2 which induces a motion of
H interchanging the points A and O will be called the "A-O-
inversion" denoted by $\Phi(A,O)$.

   We recall that in Cor. 5, IV-4, it was stated that an
h-point B is between h-points A and C if and only if B is
between A and C in the sense of the carrier of $L_h(AC)$. This
implies a property we will use, namely that if $L_h(AC)$ is
central then $S_h(AC) = S(AC)$ and if $L_h(AC)$ is non-central then
$S_h(AC) = Arc(AC)$.

Theorem 4. If an h-line t separates two h-points X and Y,
then $S_h(XY)$ intersects t.

Proof. Consider first the case in which $t = L_h(P^*Q^*)$ is central. From Th. 1, L(P*Q*) separates X and Y, so, by Axiom 4,
II-1, S(XY) intersects L(P*Q*) at a point Z. If $L_h(XY)$, as
well as t, is central, the $S(XY) = S_h(XY)$, and $Z \in S_h(XY)$
implies that Z is an h-point. Since all the h-points on
L(P*Q*) belong to t, Z is a point of t. Thus $Z \in S(XY) \cap L(P^*Q^*)$
implies $Z \in S_h(XY) \cap t$.

   If $L_h(XY)$ is non-central then S[XY] is a chord of the
circle c carrying $L_h(XY)$. Since S(XY) is interior to c, the

point Z on S(XY) is interior to c. Thus, by Th. 1, II-5,

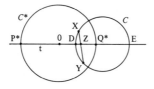

L(P*Q*) intersects c at two points D and E. Since c and
c^* are orthogonal, the points D and E are inverses with respect to c^*, (Cor. 11, IV-2), so one of them, say D, is interior to c^* and is an h-point, while E is exterior to c^*.
The segment $S_h(XY)$, which must be one of the opposite arcs,
arc(XDY) or arc(XEY), cannot be arc(XEY) because E on this
arc is not an h-point. Therefore $S_h(XY)$ is the arc(XDY) on
c. Because D on L(P*Q*) is an h-point, it is a point of t,
hence $S_h(XY)$ intersects t at D.

Next, suppose that t is non-central with a circular carrier c_t. The common diameter line of c^* and c_t intersects
c_t at an h-point A, and, by Cor. 2, $\Phi(A,0)$ is the inversion
of Th. 2 which maps t to a central h-line t_1 and maps $H_1(t)$
and $H_2(t)$ respectively onto $H_1(t_1)$ and $H_2(t_2)$. Since t separates X and Y, we may suppose X $\in H_1(t)$ and Y $\in H_2(t)$. From
$X_1 = X\Phi(A,0) \in H_1(t_1)$ and $Y_1 = Y\Phi(A,0) \in H_2(t_1)$, it follows
that t_1 separates X_1 and Y_1. By the first part of the proof,
$S_h(X_1Y_1)$ intersects t_1 at a point Z. The inversion $\Phi(A,0)$
preserves h-betweenness, (Cor. 7, IV-4), hence $S_h(X_1Y_1)$ maps
onto $S_h(XY)$. Since t_1 inverts to t, it follows that Z maps
to Z_1 in $S_h(XY) \cap t$.   □

Theorem 5. Each side of an h-line is h-convex.

Proof. Let $t = L_h(P*Q*)$, and consider two points X,Y in
the side $H_1(t)$. It must be shown that $S_h(XY) \subset H_1(t)$. If
t is non-central, then, by Th. 2, there is a motion of H
which maps t to a central h-line t_1 and maps $H_1(t)$ onto a side
$H_1(t_1)$. The points X,Y in $H_1(t)$ map to two points X_1, Y_1
in $H_1(t_1)$. Since $S_h(XY) \subset H_1(t)$ if and only if $S_h(X_1Y_1) \subset$
$H_1(t_1)$, there is no loss of generality in supposing from the
outset that t itself is central.

   If $L_h(XY)$, as well as t, is central, then $S_h(XY) = S(XY)$.
Because both the X-Y-side of $L(P*Q*)$ and H are convex (in the
ordinary sense), their intersection is convex, (Th. 13, II-1).
Their intersection is $H_1(t)$, hence $S(XY) \subset H_1(t)$, and therefore $S_h(XY) \subset H_1(t)$.

   Next, suppose that $L_h(XY)$ is non-central, with a circular
carrier c. Then $S_h(XY)$ is Arc(XY) on c. For an indirect
argument, assume that

$$S_h(XY) \not\subset H_1(t) \tag{*}$$

The (*)-assumption implies that $S_h(XY)$ intersects $H_2(t) \cup t$,
hence that one of the sets $S_h(XY) \cap H_2(t)$ or $S_h(XY) \cap t$ is
non-empty. Suppose, first, that there exists $Z \in S_h(XY) \cap$
$H_2(t)$. Then t separates X and Z and also separates Y and Z.
From Th. 4, it follows that $S_h(ZX)$ intersects t at some point
A and that $S_h(ZY)$ intersects t at some point B. By definition,
$Z \in S_h(XY)$ implies h-$<XZY>$. Thus, by the h-equivalent of
Cor. 5.2, II-1, $R_h(ZX)$ and $R_h(ZY)$ are opposite h-rays. Since
$A \in S_h(ZX) \subset R_h(ZX)$ and $B \in S_h(ZY) \subset R_h(ZY)$, it follows that $A \neq B$.
Therefore t intersects $L_h(AB)$ at two points, contradicting

Cor. 1, IV-1. Thus $S_h(XY) \cap H_2(t)$ must be empty.

Because $S_h(XY) \cap H_2(t) = \emptyset$, the (*)-assumption implies
that there exists $Z \in S_h(XY) \cap t$. If R is the non-X-Y-side of
$L(P*Q*)$, a point of $S_h(XY)$ in R would be an h-point in R and
hence a point of $H_2(t)$. Thus $S_h(XY) \cap H_2(t) = \emptyset$ implies that
$S_h(XY) \cap R = \emptyset$ hence that $Arc(XY) \cap R = \emptyset$. Since the
euclidean line $L(P*Q*)$ intersects $Arc(XY)$ at Z, if $L(P*Q*)$
were a secant of circle c then both sides of $L(P*Q*)$ would
intersect $Arc(XY)$. Therefore $Arc(XY) \cap R = \emptyset$ implies that
$L(P*Q*)$ is tangent to c at Z. But c is orthogonal to $c*$
at two points $U*$, $V*$, and $L(OU*)$ and $L(OV*)$ are the only euclidean lines through O which are tangent to c . Since neither
of these lines intersects $L_h(XY)$, neither of them intersects
$Arc(XY)$. Therefore $L(P*Q*)$ is neither of the lines $L(OU*)$,
$L(OV*)$, and hence is not tangent to c . The contradiction
shows that $Z \in S_h(XY) \cap t$ is also impossible.

Because the (*)-assumption leads to a contradiction in all
cases, it follows that $S_h(XY)$ must be contained in $H_1(t)$, hence
$H_1(t)$ is an h-convex set.   □

The last of the h-equivalents for the properties in Axiom
4 is given by the following uniqueness theorem.

Theorem 6.

If t is an h-line and if R and S are non-empty, h-convex
sets such that: (i) $H = R \cup t \cup S$; (ii) no two of the
sets R, t, S intersect; (iii) $X \in R$ and $Y \in S$ imply $S_h(XY)$
$\cap t \neq \emptyset$, then R and S are the opposite sides of t.

Proof

By hypothesis, there exists A $\in$ R and B $\in$ S . From (ii), A $\notin$ t, so we may suppose that A $\in$ H$_1$(t). If X $\in$ R $\cap$ H$_2$(t), then S$_h$(AX) $\subset$ R because R is h-convex. From Th. 5, A $\in$ H$_1$(t) and X $\in$ H$_2$(t) imply that S$_h$(AX) $\cap$ t $\neq$ $\emptyset$. Therefore R $\cap$ t $\neq$ $\emptyset$, which contradicts the hypothesis (ii). Thus X $\in$ R $\cap$ H$_2$(t) is impossible, so R $\cap$ H$_2$(t) = $\emptyset$, hence R $\subset$ H$_1$(t). The same argument shows that if B $\in$ H$_1$(t) then S $\subset$ H$_1$(t). But in that case, no point of H$_2$(t) is in R $\cup$ t $\cup$ S , which contradicts (i). Therefore B $\notin$ H$_1$(t) and, since B $\notin$ t, it follows that B $\in$ H$_2$(t). By the same argument as before, B $\in$ H$_2$(t) implies that S $\subset$ H$_2$(t).

Now suppose that there exists Z $\in$ H$_1$(t) and Z $\notin$ R . Since Z $\in$ H$_1$(t), then Z $\notin$ t and Z $\notin$ S , because S $\subset$ H$_2$(t). Therefore Z $\notin$ R $\cup$ t $\cup$ S , which contradicts (i). The contradiction implies that H$_1$(t) $\subset$ R and this, with R $\subset$ H$_1$(t), shows that R = H$_1$(t). Similarly, S = H$_2$(t).   □

Because a line preserving motion of H maps segments onto segments, Th. 6 has the following corollary.

Corollary 6.

If Γ is a line preserving motion of H and if R and S are the sides of line t, then RΓ and SΓ are the sides of line t Γ .   (Ex.)

Angles and angle interiors are now defined in H exactly as in A^2.

Definition

(Angle and angle interior in H)   An h-angle is the union

of two closed, non-collinear h-rays with a common origin. The
closed rays are the <u>arms</u> of the angle and the common origin is
the <u>vertex</u> of the angle. The angle with arms R_h [BA) and R_h [BC)
is denoted by $\star_h ABC$ or $\star_h CBA$ and the <u>interior</u> of the angle
is the set

$$In(\star_h ABC) = \text{A-side of } L_h(BC) \cap \text{ C-side of } L_h(BA).$$

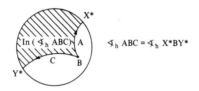

$$\sphericalangle_h ABC = \sphericalangle_h X^*BY^*$$

The definitions for "opposite h-angles", "angles of intersection of h-lines", "betweeness of h-rays", and "adjacent h-
angles" are obtained by an h-qualification of the corresponding
definitions in A^2. However, the definition of the measure, or
size, of an h-angle cannot be obtained in this automatic fashion.

If $s = L_h(P^*Q^*)$ and $t = L_h(U^*V^*)$ are two h-lines which
intersect at point B, the opposite closed h-rays at B on s and t
determine the angles of intersection of s and t. The carriers
of s and t, c_s and c_t respectively, are not tangent at B, and
the line s_1 in $F_B(c_s)$ and the line t_1 in $F_B(c_t)$ determine the
four intersection angles of c_s and c_t. Poincaré saw that he
could define the measures of the h-angles formed by s and t to
be the measures of the euclidean angles formed by s_1 and t_1. To
follow up this idea, we need a way of associating a particular
intersection angle of s and t with a particular intersection

angle of s_1 and t_1. To this end, we introduce the following
notion.

Definition

(Ray associated with an h-ray)   Associated with the open
and closed h-rays $R_h(BD)$ and $R_h[BD)$ are the open and closed
euclidean rays $R(BX)$ and $R[BX)$ such that $R[BX)$ contains
$R_h[BD)$ if $L_h(BD)$ is central and is tangent to Arc$[BD]$ if $L_h(BD)$
not central.

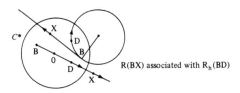

$R(BX)$ associated with $R_h(BD)$

If $L_h(BD)$ is central, then a ray associated with $R_h(BD)$ must
contain $S(BD)$ and clearly the only rays at B with this property
are $R(BD)$ and $R[BD)$.   Thus $R[BD)$ is the unique closed ray
associated with $R_h(BD)$, and if $R_h(BE)$ is opposite to $R_h(BD)$ then
$R(BE)$ is associated with $R_h(BE)$. If $L_h(BD)$ is not central then
Arc(BD) exists and Th. 2, IV-3, implies that there is a unique
closed ray $R[BX)$ tangent to Arc(BD) and hence associated with
$R_h(BD)$. If F is antipodal to B on the circle c carrying $L_h(BD)$,
then Arc(BD) is contained in a semicircle $arc_1(BF)$ on c , and,
by Th. 5, IV-3, $R(BX)$ is tangent to $arc_1(BF)$. If $R_h(BE)$ is
opposite to $R_h(BD)$, then Arc(BE) is contained in $arc_2(BF)$

opposite to arc_1(BF). The unique closed ray $\overset{\leftrightarrow}{R}$[BY) tangent to
Arc(BE), and associated with R_h(BE), is tangent to arc_2(BF) and
is therefore, by Th. 2, IV-3, a ray opposite to R(BX). Thus
ray association has the following properties.

Theorem 7.

Each h-ray has a unique closed, euclidean ray associated
with it, and the rays associated with two opposite h-rays are
opposite rays.

Corollary 7.1

If s is the carrier of two opposite h-rays at point B, then
the union of the two closed associated rays is the line in the
family $F_B(s)$.

Corollary 7.2

If two h-rays at a point B are not collinear, then their
associated rays are not collinear.

Definition

(Angle associated with an h-angle)   A euclidean angle is
associated with $\angle_h$BED if one arm of the angle is associated
with R_h(EB) and the other is associated with R_h(ED).

Corollary 7.3

Each h-angle has a unique associated angle.

Definition

(Measure of h-angles)   The _measure_ of an h-angle is the measure of its associated angle. Thus if $\measuredangle_h BED$ has the associated angle $\measuredangle XEY$ then the measure of $\measuredangle_h BED$, denoted by $\measuredangle_h BED^{\circ}$, is defined by $\measuredangle_h BED^{\circ} = \measuredangle XEY^{\circ}$. The h-angles $\measuredangle_h BED$ and $\measuredangle_h FGH$ are _congruent_ if they have the same measure, and $\measuredangle_h BED \cong \measuredangle_h FGH$ implies and is implied by $\measuredangle_h BED^{\circ} = \measuredangle_h FGH^{\circ}$.

Corollary 7.4

Opposite h-angles are congruent.

The definition of the measure of an h-angle, together with Corollary 7.3, establishes an h-equivalent to Axiom 5, II-2. That is, corresponding to each angle $\measuredangle_h BED$ there exists a number $\measuredangle_h BED^{\circ}$ between 0 and 180 which is the measure of the angle. Also, it is clear from Corollary 7.1 that if two h-lines s_1 and s_2, with carriers S_1 and S_2 , intersect at a point B, then the intersection angles of S_1 and S_2 are the angles associated with the h-angles formed by s_1 and s_2.

Axiom-h-theorems for the angle addition axiom, (Axiom 6, II-2), and the protractor axiom, (Axiom 7, II-2), can be obtained most easily by making use of motions of H induced by inversions. However, it must be shown first that the measure of h-angles is invariant in such a motion, and this property is a consequence of the next theorem.

Theorem 8.

If $C(A,r)$ is a circle orthogonal to c^{*}, and if $R_h(PX^{*})$ and $R_h(QY^{*})$ are h-rays interchanged in the inversion $\Phi_A(r)$, then the rays associated with $R_h(PX*)$ and $R_h(QY*)$ are corresponding rays in the inversion.

Proof

Let $s = L_h(X*X_1^*)$ and $t = L_h(Y*Y_1^*)$ be the lines of $R_h(PX*)$ and $R_h(QY*)$ respectively and let $R(PX)$ be associated with $R_h(PX*)$ and let $R(QY)$ be associated with $R_h(QY*)$. Finally, suppose that S and T are the carriers of s and t respectively. If S and T are both circles, then Arc(PX*) maps onto Arc(QY*). By the definition of associated rays, $R(PX)$ is tangent to Arc (PX*) and $R(QY)$ is tangent to Arc(QY*). Thus, by the definition of corresponding rays, $R(PX)$ corresponds with $R(QY)$.

Next, suppose that one of the carriers is a circle and the other is a line. We may suppose that S is the circle, and since it inverts to a line T , S is a circle through A and orthogonal to $c*$. The point X* and inverts to Y* and T is the diameter line L(OY*) of c^* . By definition, $R(PX)$ is tangent to Arc(PX*) and is therefore also tangent to the arc(PX*A) on S . By Cor. 7.1, IV-3, the image of arc(PX*A) in the inversion is the ray corresponding to $R(PX)$. Thus $R(QY*)$ corresponds with $R(PX)$, and since $R(QY*)$ contains $R_h(QY*)$ it is also the ray $R(QY)$ associated with $R_h(QY*)$.

The only case remaining, that in which S and T are both lines, occurs when $L(OA) = L(X*X_1^*) = L(Y*Y_1^*)$. Since the inversion maps $c*$ onto itself, and leaves no point of L(OA) fixed, the points X* and X_1^* are interchanged, and so $X_1^* = Y*$. The open segment $S(X*X_1^*)$, to which P and Q belong, maps onto itself and so the rays R(PX*) and $R(QX_1^*) = R(QY*)$ are oppositely directed. Thus, by Cor. 7.1, IV-3, R(QX*) and R(QY*) are corresponding rays in the inversion. But $R_h(QX*) \subset R(QX*)$ implies that

$R(QX^*) = R(QX)$, and $R_h(QY^*) \subset R(QY^*)$ implies that $R(QY^*) = R(QY)$.
Therefore the associated rays are corresponding in the inversion. □

Corollary 8.

The motion of $\aleph$ induced by $\Phi_A(r)$ preserves the measure of h-angles.

Proof

If the inversion maps $\divideontimes_h DEF$ to $\divideontimes_h D'E'F'$, it follows from Th. 6 that the angle associated with $\divideontimes_h DEF$ and that associated with $\divideontimes_h D'E'F'$ are corresponding angles and, by Th. 12, IV-3, these corresponding angles are congruent. By definition, the measure of an h-angle is the measure of its associated angle, and so the angles $\divideontimes_h DEF$ and $\divideontimes_h D'E'F'$ are congruent. □

We turn now to the h-equivalent of the angle addition axiom (Axiom 6, II-2).

Theorem 9.

If $R_h(BD) \subset In(\divideontimes_h ABC)$ then
$$\divideontimes_h ABD \;^{\mathrm{o}} + \divideontimes_h DBD \;^{\mathrm{o}} = \divideontimes_h ABC \;^{\mathrm{o}}.$$

Proof

If A is O, the associated angles of $\divideontimes_h AOD$, $\divideontimes_h DOC$, and $\divideontimes_h AOC$ are $\divideontimes AOD$, $\divideontimes DOC$, and $\divideontimes AOC$ respectively. The h-interior of $\divideontimes_h AOC$ is a subset of $In(\divideontimes AOC)$. Thus $R_h(OD) \subset In(\divideontimes_h AOC)$ implies that $R(OD) \subset In(\divideontimes AOC)$. From the angle addition axiom, it follows that
$$\divideontimes AOD^{\mathrm{o}} + \divideontimes DOC \;^{\mathrm{o}} = \divideontimes AOC \;^{\mathrm{o}}. \tag{1}$$
The equalities $\divideontimes AOD^{\mathrm{o}} = \divideontimes_h AOD \;^{\mathrm{o}}$, $\divideontimes DOC \;^{\mathrm{o}} = \divideontimes_h DOC \;^{\mathrm{o}}$, and

∢AOC° = ∢$_h$AOC°, together with (1) imply that

$$∢_hAOD^\circ + ∢_hDOX^\circ = ∢_hAOC^\circ. \qquad (2)$$

If $A \neq 0$, consider the inversion $\Phi(B,0)$, mapping B to 0. Because the motion of H induced by $\Phi(B,0)$ maps h-lines onto h-lines, it follows from Cor. 4 that the interior of an h-angle maps onto the interior of the image h-angle. Denoting images by primes, with $B' = 0$, it follows that $R_h(BD) \subset In(∢_hABC)$ implies that $R_h(OD') \subset In(∢_hA'OC')$. By the same argument as in the preceding paragraph, $R_h(OD) \subset In(∢_hA'OC')$ implies that

$$∢_hA'OD'^\circ + ∢_hD'OC'^\circ = ∢_hA'OC'^\circ. \qquad (3)$$

Since $\Phi(B,0)$ preserves the measure of h-angles, $∢_hA'OD' \cong ∢_hABD$, $∢_hD'OC' \cong ∢_hDBC$, and $∢_hA'PC' \cong ∢_hABD$. These congruences, together with (3), imply that

$$∢_hABD^\circ + ∢_hDBC^\circ = ∢_hABC^\circ. \qquad (4) \;\square$$

We obtained an h-equivalent to Axiom 5 by defining the number that is the measure of an h-angle. Similarly, we can obtain an h-equivalent to the protractor axiom, (Axiom 7, II-2), by defining the number that plays the role of an h-ray coordinate in the h-space formulation of the axiom.

Theorem 10.

If $R_h[A\,B)$ is a closed ray in the edge line t of an open half-plane $H_1(t)$, and if the ray coordinate x of an open ray $R_h(AX) \subset H_1(t)$ is defined by $x = ∢_hBAX^\circ$, then the correspondence $R_h(AX) \longleftrightarrow x$, $R_h(AX) \subset H_1(t)$, is a one-to-one correspondence between the open rays, with origin A, in $H_1(t)$ and the set of real numbers between 0 and 180.

Proof

Suppose, first, that t is a central h-line and that

$R_h[AB) = R_h[OB)$. If $R_h(OX) \subset H_1(t)$, then $x = \measuredangle_h BOX^\circ = \measuredangle BOX^\circ$, and so $0 < x < 180$. Moreover, if $R_h(OX)$ and $R_h(OY)$ are distinct rays in $H_1(t)$, then $R_h(OX) \neq R_h(OY)$ implies that $R(OX) \neq R(OY)$. Therefore $\measuredangle BOX^\circ \neq \measuredangle BOY^\circ$, by Axiom 6, II-2, hence $\measuredangle_h BOX^\circ \neq \measuredangle_h BOY^\circ$, and so $x \neq y$. Thus the matching of the rays with their coordinates is at least a one-to-one correspondence of the rays $R_h(AX)$ in $H_1(t)$ with a subset of the numbers between 0 and 180. But if x is a real number between 0 and 180, then by Axiom 6, II-2, there exists a unique ray $R(OX)$ in the euclidean $H_1(t)$-side of $L(OA)$ such that $\measuredangle BOX^\circ = x$. This ray intersects the c^*-boundary of $H_1(t)$ at a point X^*, and $\measuredangle BOX$ is the associated angle of $\measuredangle_h BOX^*$. Thus $R_h(OX^*)$ is a ray in $H_1(t)$ with ray coordinate x. Therefore, $R_h(OX) \longleftrightarrow x$, $R_h(OX) \subset H_1(t)$, is a one-to-one correspondence between the set of open rays, with origin A, in $H_1(t)$ and the set of all real numbers between 0 and 180.

Now consider the correspondence $R_h(AX) \longleftrightarrow x (= \measuredangle_h BAX^\circ)$, $R_h(AX) \subset H_1(t)$, where t is any line in H, but $A \neq 0$. In the motion of H induced by the inversion $\Phi(A,0)$, t maps to a central h-line t', $R_h[AB)$ in t maps to $R_h[A'B') = R_h[OB')$ in t', and $H_1(t)$ maps onto $H_1(t')$. The correspondence $R_h(AX) \longleftrightarrow R_h(OX')$, $R_h(AX) \subset H_1(t)$, is a one-to-one correspondence of the open rays at A in $H_1(t)$ with the open rays at 0 in $H_1(t')$ As already shown, the correspondence $R_h(OX') \longleftrightarrow \measuredangle_h B'OX'^\circ$, $R_h(OX') \subset H_1(t')$, is a one-to-one correspondence between the set of open rays at 0 in $H_1(t')$ and the set of real numbers between 0 and 180. Therefore, $R_h(AX) \longleftrightarrow \measuredangle_h B'OX'^\circ$, $R_h(AX) \subset H_1(t)$, is a one-to-one correspondence between the set of open rays at A in $H_1(t)$ and the set of numbers between 0 and 180. But since $\Phi(A,0)$ preserves the measure of h-angles, (Cor. 6), $\measuredangle_h B'OX'^\circ = \measuredangle_h BAX^\circ = x$. Therefore the correspondence $R_h(AX) \longleftrightarrow \measuredangle_h B'OX'^\circ$, $R_h(AX) \subset H_1(t)$ is the correspondence $R_h(AX) \longleftrightarrow x$, $R_h(AX) \subset H_1(t)$, and so this correspondence

is a one-to-one matching of the open rays at A in $H_1(t)$ with
the real numbers between 0 and 180.   □

Exercises - Section 5

1. Prove Th. 1.

2. Prove Cor. 2.

3. Prove Th. 3

4. Prove that the interior to a euclidean circle is a convex set.

5. Prove Cor. 6.

6. Give definitions in H for: (i) opposite h-angles; (ii) betweeness of h-rays; (iii) adjacent h-angles.

7. Prove that every euclidean ray whose origin is an h-point is the associated ray of a unique, closed h-ray.

8. Given ≮XPY such that P is an h-point and $≮XPY^o = 50^o$, draw and discuss a diagram which illustrates the angle $≮_h X*PY*$ of measure 50° and such that ≮XPY is its associated angle.

9. How does Th. 10 imply that if $≮_h ABC^o = ≮_h ABD^o$ and if C and D are in one side of $L_h(AB)$ then $R_h(BC) = R_h(BD)$?

10. The h-equivalent to Th. 15, II-2, implies that there exists a unique ray $R_h(AZ*)$ which bisects $≮_h X*AY*$. If $A \neq 0$, and if B is the inverse of A with respect to $c*$, let u be the euclidean line which is the perpendicular bisector of $S[AB]$. If u and $L(X*Y*)$ intersect at E, and if $C(E,r)$ is orthogonal to $c*$, explain why one intersection of $C(E,r)$ and $c*$ is the point $Z*$. How is the bisector ray $R_h(AZ*)$ determined when u is parallel to $L(X*Y*)$?

11. By the h-equivalent to Th. 18, II-2, if A is a point of the h-line s then there is exactly one h-line t which is perpendicular to s at A. Explain a construction for t.

12.   If $s = L_h(X*Y*)$ and $t = L_h(U*V*)$ are hyperparallel, and
      if the lines $L(X*Y*)$ and $L(U*V*)$ intersect at E, why does
      the circle $C(E,r)$ which is orthogonal to $c*$ carry the
      h-line which is the common perpendicular to s and t?
      How is the common perpendicular to s and t determined when
      $L(X*Y*)$ and $L(U*V*)$ are euclidean parallels?

Section 6. Triangle Congruence in the Model, the Consistency of Hyperbolic Geometry

If no h-line contains the three h-points A, B, C, then
in pairs they determine three segments $S_h[AB]$, $S_h[BC]$, and
$S_h[CA]$ whose union is the h-triangle $\triangle_h ABC$. All the concepts
for triangles in A^2 have h-counterparts for triangles in H and,
in particular, a correspondence ABC < - > DEF is a congruence
of $\triangle_h ABC$ and $\triangle_h DEF$, denoted $\triangle_h ABC \cong \triangle_h DEF$ if the corresponding segments are congruent and the corresponding angles are
congruent. The last of the axioms for absolute geometry, Axiom
8, II-3, was the side-angle-side congruence condition for triangles. In this section we want to establish the h-counterpart
to this axiom and to discuss what the complete list of axiom-
h-theorems implies about hyperbolic geometry.

Suppose that ABC < - > DEF is a correspondence of $\triangle_h ABC$
and $\triangle_h DEF$ in which two sides and an included angle in one
triangle are congruent to the corresponding sides and included
angle in the other, say, for instance, that $S_h[BA] \cong S_h[ED]$,
$\angle_h B \cong \angle_h E$, and $S_h[BC] \cong S_h[EF]$. One rather natural idea for
a proof that the other corresponding parts are also congruent
is to use mappings. If Γ is any motion of H which maps h-
lines onto h-lines and preserves the measure of h-angles, then
clearly each triangle in H is automatically congruent to its

image triangle in the mapping. Thus if it can be shown that
the given information about the triangles implies the existencê
of such a motion in which A,B,C map to D,E,F respectively, then
the desired theorem will be established.

In implementing the plan just described, the following
agreements will be helpful.

Conventions:

If t is a central h-line, then "the reflection in t" will
refer to the reflection of E^2 in the diameter line of c^* which
carries t. If t is non-central, "the reflection in t" will
refer to the inversion in the circle c which carries t and which
maps Cp(A) onto itself, if A is the center of c . In either
case, we will use the notation Γ_t to represent the reflection.

If t is a central h-line, the reflection Γ_t leaves c^*
invariant and hence, by Th. 6, IV-4, maps h-lines onto h-lines.
Because Γ_t is a motion of E^2 , it preserves the tangency of
a ray and an arc. Since Γ_t also preserves set inclusions, it
follows that the ray associated with an h-ray maps onto the ray
associated with the image h-ray. Thus Γ_t maps the associated
angle of an h-angle onto the associated angle of the image h-
angle. Because angle measure is invariant, the associated angles
are congruent, and therefore the h-angles are congruent. If
t is non-central, it follows from Th. 12, IV-3, that Γ_t maps
h-lines onto h-lines and preserves the measure of h-angles.
Thus, whether t is central or non-central, Γ_t has the following
properties.

Theorem 1.

If t is an h-line, the motion of H induced by the reflec-

tion Γ_t maps h-lines onto h-lines and preserves the measure
of h-angles. The fixed h-points are the points of t, the sides
of t are interchanged, and the invariant h-lines are t and the
lines perpendicular to t.

Corollary 1.

If the reflection Γ_t maps the vertices A,B,C of $\triangle_h ABC$
to A', B', C' respectively, then $\triangle_h ABC \cong \triangle_h A'B'C'$.

We can simplify the proof of an h-equivalent to Axiom 8
by the use of two reflection properties which we state as
lemmas.

Lemma 1.

If A and B are distinct h-points, there exists an h-
reflection Γ_t which maps A onto B.

Proof

By Th. 9, II-1, there exists a midpoint M to S_h [AB] and,
by Th. 18, II-2, there exists an h-line t which is perpendicular
to L_h (AB) at M. By Th. 1, the reflection Γ_t leaves M fixed,
leaves t invariant, and interchanges the sides of t. Thus R_h (MA)
maps onto R_h (MB). The point A' = AΓ_t is therefore on R_h (MB),
and since h(A',M) = h(A,M) = h(B,M), it follows from Th. 6,
II-1, that A' = B.   □

Lemma 2.

If R_h (AB) and R_h (AC) are distinct rays, there exists an
h-reflection Γ_t which maps R_h (AB) onto R_h (AC).

Proof

If $R_h(AB)$ and $R_h(AC)$ are collinear then clearly they are
opposite rays. There exists a line t which is perpendicular to
$L_h(AB)$ at A, (Th. 18, II-2), and by Th. 1, Γ_t maps $R_h(AB)$
onto $R_h(AC)$. If $R_h(AB)$ and $R_h(AC)$ are not collinear, then by
Th. 15, II-2, there exists a ray $R_h(AD)$ which bisects $\not{\times}_h BAC$. If
t is the h-line of $R_h(AD)$, Th. 1 implies that Γ_u maps $R_h(AB)$ to
$R_h(AB)$ to $R_h(AB')$ in the C-side of t. From $\not{\times}_h DAB^o = \not{\times}_h DAC^o$
and $\not{\times}_h DAB^o = \not{\times}_h DAB'^o$ it follows that $\not{\times}_h DAB'^o = \not{\times}_h DAC^o$, and this
equality, by Th. 10, IV-5, implies that $R_h(AB') = R_h(AC)$.   □

From the definition of triangle congruence it follows that
if $\triangle ABC \cong \triangle DEF$ and $\triangle DEF \cong \triangle GHI$ then $\triangle ABC \cong \triangle GHI$.
We make use of the transitivity in the proof of the following
h-equivalent to Axiom 8, II-3.

Theorem 2.

If a correspondence of two h-triangles, or of an h-triangle
with itself, is such that two sides and the angle between them
are respectively congruent to the corresponding two sides and
the angle between them, the correspondence is a congruence of the
triangles.

Proof

Let ABC <-> DEF denote a correspondence of $\triangle_h ABC$ and
$\triangle_h DEF$ in which $S_h[BA] \cong S_h[ED]$, $\not{\times}_h ABC \cong \not{\times}_h DEF$, and $S_h[BC] \cong S_h[EF]$.
If $B \neq E$, let Γ_1 denote the h-line reflection which maps B to E,
(Lemma 1), and if $B = E$ let Γ_1 denote the identity mapping I on
E^2. In either case, from Cor. 1 or the properties of I, it
follows that Γ_1 maps A, B, C, to points A_1, B_1, C_1, such

that $B_1 = E$ and $\Delta_h ABC \cong \Delta_h A_1 B_1 C_1 = \Delta_h A_1 E C_1$. Now, if $R_h(EA_1)$
$\neq R_h(ED)$ let Γ_2 denote the h-line reflection which maps $R_h(EA_1)$
onto $R_h(ED)$, (Lemma 2), and if $R_h(EA_1) = R_h(ED)$ let Γ_2 denote
the identity mapping I. In either case, Γ_2 maps A_1, E, C_1 to the
points A_2, E, C_2 respectively such that $\Delta_h A_1 E C_1 \cong \Delta_h A_2 E C_2$,
and $R_h(EA_2) = R_h(ED)$. Finally, if $t = L_h(ED) = L_h(EC_2)$ separates C_2 and F, let $\Gamma_3 = \Gamma_t$, otherwise let $\Gamma_3 = I$. In either
case, Γ_3 maps A_2, E, C_2 to A_2, E, C_3 respectively such that
$\Delta_h A_2 E C_2 \cong \Delta_h A_2 E C_3$, and $R_h(EC_3)$ is in the F-side of t.
Because the congruence of triangles is transitive, the sequence
of congruences implies that

$$\Delta_h ABC \cong \Delta_h A_2 E C_3. \tag{1}$$

By hypothesis, $S_h[BA] \cong S_h[ED]$ and, from (1), $S_h[BA] \cong$
$S_h[EA_2]$. Therefore, $h(E,D) = h(E,A_2)$ and since $R_h(ED) =$
$R_h(EA_2)$ it follows from Th. 6, II-1, that $A_2 = D$. Thus (1) can
also be expressed as

$$\Delta_h ABC \cong \Delta_h DEC_3. \tag{2}$$

By hypothesis, $\not\supset_h ABC^\circ = \not\supset_h DEF^\circ$ and, from (2), $\not\supset_h ABC^\circ =$
$\not\supset_h DEC_3^\circ.$ Therefore $\not\supset_h DEF^\circ = \not\supset_h DEC_3^\circ$ and, since $R_h(EF)$ and
$R_h(EC3)$ are in the same side of $L_h(ED)$, Th. 10, IV-5, implies
that $R_h(EF) = R_h(EC_3)$. By hypothesis, $h(B,C) = h(E,F)$ and,
from (2), $h(B,C) = h(E,C_3)$. Therefore $h(E,F) = h(E,C_3)$ and so,
by Th. 6, II-1, $C_3 = F$. Thus (2) is equivalent to

$$\Delta_h ABC \cong \Delta_h DEF, \tag{3}$$

and so the correspondence ABC $\leftrightarrow$ DEF is a congruence. $\square$

With Th. 2, h-equivalents have been established for all
the axioms of absolute geometry. Thus every theorem in Chapter II
has an h-equivalent. In particular, Th. 6, III-2 implies that
the sum of the lengths of each two sides of an h-triangle is
greater than the length of the third side. From this it follows

that if A, B, C are any three h-points, then $h(A,B) + h(B,C) \geqq$
$h(A,C)$ and the equality holds if and only if h-<ABC>. Therefore
every motion of H preserves h-betweenness of points, because it
preserves distance, and therefore every motion of H maps h-lines
onto h-lines. Also, if Γ is a motion of H and if A, B, C are
non-collinear, then each two numbers in the set { $h(A,B)$, $h(B,C)$,
$h(C,A)$ } have a sum greater than the third number in the set.
Because Γ is a one-to-one mapping, the image points A', B', C'
are distinct and { $h(A',B')$, $h(B',C')$, $h(C',A')$ } is the same
set as before. Thus the sum of each two of these numbers is
greater than the third, and so A', B' and C' are non-collinear
and $\triangle_h A'B'C'$ exists. From the side-side-side triangle congruence theorem, (Th. 8, II-3), it follows that $\triangle_h A'B'C' \cong$
$\triangle_h ABC$. Therefore $\not{\succ}_h A'B'C' = \not{\succ}_h ABC$. Thus the following
properties hold for all motions of H and not just for line
reflections.

Theorem 3.

Every motion of H maps h-lines onto h-lines and preserves
the measure of h-angles.

It is, of course, a formality to show that there is an h-
equivalent to the hyperbolic axiom, (Axiom 9, III-1). In fact,
if the h-point P is not on the line $t = L_h(X*Y*)$, neither of
the lines $L_h(PX*) = L_h(X*X_1^*)$ and $L_h(PY*) = L_h(Y*Y_1^*)$ intersects
t. These lines are the two parallels to t at P and $\not{\succ}_h X*PY*$ is
the fan angle of P and t and contains t in its interior. The
h-lines which subdivide the opposite angles $\not{\succ}_h X*PY*$ and
$\not{\succ}_h X_1^* PY_1^*$ are the intersectors of t in the pencil $P_h(P)$, and

the lines which subdivide the opposite angles $\not\lessgtr_h X*PY_1^*$ and
$\not\lessgtr_h X_1^* PY*$ are the hyperparallels to t in the pencil $P_h(P)$.

Because h-equivalents exist for the axioms of hyperbolic
geometry, as well as for those of absolute geometry, all the
theorems in Ch. III have h-equivalents and, in fact, every
theorem of hyperbolic plane geometry has an h-equivalent. The
h-geometry is hyperbolic plane geometry and a further investigation of the subject, beyond the content of Ch. III, could be
carried on in either H^2 or in H .

An important feature of the model H is that it can be
viewed in two different ways. From one point of view, as just
noted, the model is a representation of hyperbolic geometry.
But from another point of view, it is simply a special subsystem of euclidean geometry. All the objects in H are objects
in E^2 and every h-theorem is a theorem of euclidean geometry
about those objects. These two aspects of the model imply that
if the axiom system of hyperbolic geometry is logically inconsistent then so is the axiom system of euclidean geometry. For
suppose that there exists some proposition P about objects in
H such that P and its denial, say $\bar{P}$, are both implied by the
axioms for H^2 . Then because of the equivalence of the geometry in H^2 and the geometry in H , the h-quivalents to P and

to $\overline{P}$, say P_h and $\overline{P}_h$ must both be theorems in the h-geometry.
But h-theorems are theorems of euclidean geometry. And a
proposition P_h in euclidean geometry and its denial $\overline{P}_h$ cannot
both be theorems unless the axiom system of euclidean geometry
is logically inconsistent. Thus the equivalence of the geometry of H^2 and the geometry of H implies that hyperbolic plane
geometry and euclidean plane geometry are both logically consistent or else both are logically inconsistent. As a consequence,
of course, no logical refutation of hyperbolic geometry is
possible unless there is a corresponding refutation of euclidean
geometry.

We conclude with a few informal and descriptive observations
about the model. It is interesting, for example, that one can
literally see the defect of certain h-triangles. Let t denote
an h-line whose circular carrier c has center A and let F be the
intersection of t and L(OA). If P on t is not F, the reflection
in $u = L_h(OF)$, namely Γ_u, maps t onto itself and maps P to a

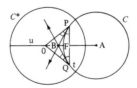

point Q such that $\Delta_h POQ$ is isosceles with base $S_h[\,PQ\,]$. The
tangent rays to Arc(PQ) at P and at Q intersect at a point B
interior to $\Delta_h OPQ$. The congruent angles, ⍟OPB and ⍟OQB, are
the respective associated angles of the congruent base angles,

$\angle_h OPQ$ and $\angle_h OQP$, in $\triangle_h OPQ$. The associated angle of $\angle_h POQ$ is $\angle POQ$ and so the angle sum of $\triangle_h POQ$ is $\angle POQ^\circ + 2 \angle OPB^\circ$. The angle sum of $\triangle POQ$ is $\angle POQ^\circ + 2 \angle OPQ^\circ = 180^\circ$. Thus the defect of $\triangle_h POQ$ is $2(\angle OPQ^\circ - \angle OPB^\circ) = 2\angle BPQ^\circ$.

In the figure just discussed, the fact that $\triangle POQ$ and $\triangle_h POQ$ are both isosceles is a particular consequence of the property that an h-circle with center O and radius r, denoted $C_h(O,r)$, is also a euclidean circle $C(O,k)$, though $r \neq k$. If X is an h-point distinct from O, the $\Phi(O,X)$ inversion which maps O to X preserves h-distance and hence maps an h-circle $C_h(O,r)$ to an h-circle $C_h(X,r)$. Because $C_h(O,r)$ is a euclidean circle which does not pass through the center of the inversion $\Phi(O,X)$ it inverts to a circle. Thus $C_h(X,r)$ is a euclidean circle, though the euclidean center is not X but a point of $S(OX)$. The argument is reversible, and every euclidean circle which is interior to $c*$ is also an h-circle. Each of the concentric h-circles with center X is orthogonal to all the h-lines in the pencil $P_h(X)$.

For each point X* on $c*$, the ray $R_h(OX*)$ determines the

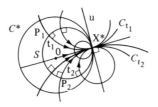

parallel family of lines $F[R_h(OX*)]$. If u is the line tangent to $c*$ at X*, each circle with a center on u and passing through

X* carries an h-line parallel to $L_h(OX^*)$ in the direction $R_h(OX^*)$, and these h-lines, together with $L_h(OX^*)$, constitute the parallel family $F[R_h(OX^*)]$. If t_1 and t_2 are two lines in this family, with carriers c_{t_1} and c_{t_2} then a circle s which is internally tangent to c^* at X^* intersects t_1 and t_2 at points P_1 and P_2 respectively. Because s is orthogonal to c_{t_1} and c_{t_2} at X*, it is orthogonal to t_1 and t_2 at P_1 and P_2 respectively. Thus s is a curve in H which is orthogonal to all the lines in $F[R_h(OX^*)]$. It is not difficult to verify that s is a limit circle and that $F[R_h(OX^*)]$ is its family of radial lines. The parallel rays $R_h[P_1X^*)$ and $R_h[P_2X^*)$ are radial rays, and the h-biangle $(X^* -P_1P_2-X^*)_h$ is isosceles. The circles which are internally tangent to c at X* form the family of limit circles co-radial with s.

Now consider a non-central h-line $b = L_h(X^*Y^*)$ with a carrier c_b whose center is B. The hyperparallel family $F_h(b)$ consists of the h-lines perpendicular to b, and these may be found as follows. If point Z on $u = L(X^*Y^*)$ is exterior to c^*, there exists a line $L(ZT^*)$ tangent to c^* at T*, and the circle c_z with center Z and radius $d(Z,T^*)$ is orthogonal to c^*. Each such circle c_z is the carrier of an h-line t in the family $F_h(b)$. First, because c_z is orthogonal to c^*, the power of Z with respect to c^* is $d(Z,T^*)^2$, (Cor. 7, IV-2). But, by Cor. 9, IV-2, the power of Z with respect to every circle through X* and Y*

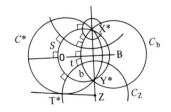

is $d(Z,X*)d(Z,Y*)$. Since $c*$ passes through X* and Y*, it follows that $d(Z,T*)^2 = d(Z,X*)d(Z,Y*)$. Thus c_z is orthogonal to every circle through X* and Y*, hence to c_b and therefore $t \perp b$ implies $t \in F_h(b)$. The h-line carried by L(OB) together with the h-lines carried by the circles c_z, just defined, form the hyperparallel family $F_h(b)$.

Now consider a circle s which passes through X* and Y* but is not orthogonal to $c*$. There exists $arc_1(X*Y*)$ on s and interior to $c*$ and this arc is a curve in H. Because c_z is orthogonal to every circle through X* and Y* it is orthogonal to s and so $arc_1(X*Y*)$ is orthogonal to line t and to every line in the family $F_h(b)$. As this suggests, $arc_1(X*Y*)$ is an equi-distant curve with base line b. The reflection r_b leaves b invariant and maps s to a circle s , (unless B is on s), which passes through X* and Y*. The arc of X* and Y* on s and interior to $c*$ is an equidistant curve at the same distance from b as $arc_1(X*Y*)$, but in the opposite side of b. The circles through X* and Y* which are not orthogonal to $c*$ intersect H in a family of open arcs and these, together with S(X*Y*), form the family of equidistant curves with baseline b and co-radial with $arc_1(X*Y*)$.

The relation of euclidean circles to the model geometry can be summarized as follows. If a euclidean circle c intersects H the intersection is: (i) an h-line if c is orthogonal to $c*$; (ii) an h-circle if c is contained in H; (iii) a limit circle if c is internally tangent to $c*$; (iv) an equidistant curve if c and $c*$ intersect twice but are not orthogonal.

Exercises - Section 6.

1. If $P \neq 0$, and $h(0,P) = r$, the circle $C_h(0,r)$ which is
    defined to be the set $\{X: h(0,X) = r\}$, passes through
    P. Alternately, (as in H^2), $C_h(0,r)$ is also the set
    $\{P\Gamma_t: t \in P_h(0)\}$. The euclidean circle $C(0,k)$ which pass-
    es through P is also the set $\{P\Gamma_t: t \in P(0)\}$. How do
    the definitions in terms of Γ_t imply that $C_h(0,r) = C(0,k)$?

2. Let s denote a circle which is contained in $c^* \cup H$ and is
    tangent to c^* at X*. If P and Q are h-points on s and if
    the line u tangent to c^* at X* intersects L(PQ) at E, the
    circle with center E and which passes through X* is the
    carrier of an h-line v. Show that the reflection Γ_v maps
    s onto itself and maps $R_h(PX^*)$ onto $R_h(QX^*)$, thus show
    that Q belongs to the limit circle $LC[R_h(PX^*)]$. If u $\parallel$
    L(PQ), what is the line v in $F[R_h(PX^*)]$ such that Γ_v maps
    s onto itself and maps $R_h(PX^*)$ onto $R_h(QX^*)$?

3. Let $b = L_h(X^*Y^*)$ be a central line in H, and let s be a
    circle through X* and Y* with $arc_1(X^*Y^*) = s \cap H$. The h-
    line v which is perpendicular to b at 0 intersects
    $arc_1(X^*Y^*)$ at a point P. Define $h(0,P) = k$. If X on
    $arc_1(X^*Y^*)$ is not on P, let Y be its inverse with respect to c.
    The perpendicular bisector of S[XY] intersects L(X*Y*) at
    a point E, and the circle c with center E and passing
    through X intersects b at F, the foot of X in b, and so
    $h(X,b) = h(X,F)$. The circle c also intersects L(X*Y*) at
    a second point G. If C(G,r) is orthogonal to c^*, show that
    $\Phi_G(r)$ induces a motion of H in which X maps to P and F maps
    to 0. Thus show that $h(X,b) = k$ and hence that $arc_1(X^*Y^*)$
    is an equidistant curve with baseline b.

4. If the euclidean line $t = L(X^*Y^*)$ does not pass through 0,

Exercises - Section 6

show that there exists a circular inversion which induces
a motion of H and which maps t to a circle through the
endpoints of a diameter of c^*. Thus, using Ex. 3, show
that every euclidean line which intersects H, but does
not pass through O, intersects H in an equidistant curve.
How could you find the base line to the equidistant curve
$S(X*Y*)$?

Appendix. Distance Geometries

Introduction: In a famous mathematical paper, written in 1854, Bernhard Riemann (1826-1866) presented a new and revolutionary point of view about the foundations of geometry and the origin of different geometries. The following is a very rough description of a central idea in this paper. Riemann argued that if one specified a rule for calculating small distances in a space, a rule called the "line element" in calculus, then one could, by a process called "integration", determine the lengths of various paths connecting two points A and B in the space. The shortest of such paths (a notion we will accept as intuitive) is a geodesic connection of A and B. These geodesics have the character of the 'lines' of the space and determine its geometry, and different line elements produce different geometries. These are now called "Riemannian geometries" and include euclidean and non-euclidean geometries as special cases. In particular, the elliptic plane geometry mentioned in I-5 was found to be a Riemannian geometry. Later, assumptions were devised to produce this geometry from the axiomatic point of view.

An accurate explanation of Riemann's ideas involves mathematics beyond the scope of this book. However, a concept stemming from Riemann's influence, the notion of a metric space, can be defined and illustrated in terms of elementary mathematics, and provides a new perspective of various distance geometries. Our object in this appendix is to convey, briefly and informally, some idea of that perspective by discussing a few special topics.

Topic I. Metric Space and Metric Geometry

The most basic properties one associates with distance are used in the following definition to characterize a distance function or metric.

<u>Definition</u>. (Metric, metric space) If s is a set of elements {A, B, C, ...} and if m(A,B) is a real number associated with the elements A,B then m is a <u>metric</u> (or distance) for s if it has the following properties for all A, B, C, in s :

    (i)   (symmetry) m(A,B) = m(B,A);

    (ii) (postivity) m(A,B) > 0 if A ≠ B;

    (iii) (zero-identity) m(A,B) = 0 if A =B;

    (iv) (triangle inequality) m(A,B) + m(B,C) $\geqq$ m(A,C).

The set s, with metric m, is a <u>metric space</u>, denoted by (s,m).

The properties (ii) and (iii) above are sometimes combined in the single statement that m(A,B) $\geqq$ 0 with equality if and only if A = B.

If the elements of a metric space (s, m) are, by agreement, called the "points" of the space, then any concept in ordinary geometry which depends only on the notion of points, and the distance between points, has a natural counterpart in the space (s,m). The following are some examples.

<u>Definition</u>. (Sphere) In the metric space (s,m), the <u>sphere</u> with center A and radius r > 0 is the set

$$S_m(A,r) = \{X: m(A,X) = r \}.$$

<u>Definition</u>. (Betweeness of points, segments) In the metric space (s,m), point B is <u>between</u> points A and C, denoted by m-<ABC> or m-<CBA>, if A,B,C are three points and

$$m(A,B) + m(B,C) = m(A,C).$$

The opened and closed segments of two points A,B are respectively the sets

$S_m(AB) = \{X:m\text{-}<AXB>\}$, $S_m[AB] = \{ X: X=A, X=B, \text{ or } m\text{-}<AXB>\}$.

Definition. (Line of two points) In a metric space (s,m)
the line of two points A B is the set

$L_m(AB) = \{ X: X=A, X=B, m\text{-}<AXB>, m\text{-}<ABX>, \text{ or } m\text{-}<XAB> \}$.

Congruence of sets can also be extended in the following
rather natural way.

Definition. (Congruent sets, congruent spaces) If R_1 and
R_2 are subsets of the metric space (s_1,m_1) and (s_2,m_2) respectively, R_1 is congruent to R_2 if there exists a mapping Γ of
R_1 onto R_2 such that $m_1(X,Y) = m_2(X\Gamma,Y\Gamma)$ for all X,Y, in R_1.
In particular, if $s_1 \cong s_2$, then (s_1,m_1) and (s_2,m_2) are
congruent spaces.

These definitions provide a logical meaning for the segments, lines, and spheres of an abstract metric space, and for
congruence of sets in the space. However, the notion of a metric space is so general, and imposes such weak restrictions,
that a particular metric space may be little more than a logical curiosity. For example, consider any collection of objects,
say all the lines and circles in the euclidean plane. Name
this collection a "space s" and name the elements of s
"s-points". If the distance m between s-points A,B is defined
to be 1 if A ≠ B and to be zero if A = B, then (s,m) is a metric
space. In this space, no S-point is between two others, so
there are no open segments. The line $L_m(AB)$ is just the pair of
points A,B, and the whole space s is a sphere which has every

point of the space as a center.

In contrast to the example above, if s is taken to be the set of points interior to a circle in the euclidean plane, and if m(X,Y) is defined to be the Poincaré distance h(X,Y), it was shown in Ch. IV that the space (s,m) is the hyperbolic plane. Thus, whether or not a metric space (s,m) has an interesting geometry depends upon the nature of the space s and the .metric m.

The space of all real numbers is commonly called a "vector 1-space" and denoted by v^1. Similarly, the space of all ordered pairs of real numbers is the vector 2-space v^2, and the space of all ordered triples of real numbers is the vector 3-space v^3. In v^1, if d(x,y) is defined by d(x,y) = $|$ x-y $|$, then (v^1,d) is a metric space which is commonly called a "euclidean 1-space" and denoted by E^1. In v^2, if the distance between the ordered pairs X = (x_1,y_1) and Y = (x_2,y_2) is defined by

$$d(X,Y) = [(x_1-x_2)^2 + (y_1-y_2)^2]^{1/2},$$

then (v^2,d) is a metric space which is called a "euclidean 2-space" and denoted by E^2. If the Playfair axiom (or some equivalent) is added to the axioms for absolute plane geometry, the resulting euclidean plane can be shown to be congruent to (v^2,d).

Convention. Throughout this appendix, the letter "d" will be reserved for the euclidean metric, E^2 will refer to the euclidean plane, and E^3 will refer to the space of euclidean solid geometry.

In all the metric spaces (s,m) which we will consider, the space s will be v^2 , E^2 , or $\bar{E}^3$, or a subset of one of these spaces.

It is interesting to observe that the Birkhoff ruler axiom

from our present point of view, is essentially the assumption

that the sets of points called lines in A^2 are sets which are

congruent to the space $B^1 = (V^1, d)$. Since such lines exist

in A^2, in E^2, and in H^2, we can, somewhat arbitrarily, call any

set in a space (S,m) a "natural line" if it is congruent to

(V^1, d).

Topic I Exercises.

1. If $R_1 \subset (S_1, m_1)$ and $R_2 \subset (S_2, m_2)$ and if Γ is a mapping
    of R_1 onto R_2 such that $m_1(X,Y) = m_2(X\Gamma, Y\Gamma)$ for all
    X, Y in R_1, explain why Γ must be a one to one mapping.

2. Show that if m is a metric for a space S and if m' is
    defined by $m'(A,B) = [m(A,B)]^{1/2}$ for all A, B in S, then
    (S, m') is a metric space. Prove that in (S, m') there is
    no point that is between two others.

3. Show that if (S,m) is a metric space, and $R \subset S$, then
    (R,m) is a metric space.

4. Show that if k is a positive number, and if $m(A,B)$ is a
    metric for a space S, then $m'(A,B) = km(A,B)$ is also a
    metric for S.

5. From the metric space point of view, the familiar euclidean
    cartesian plane, with perpendicular x and y axes, is the
    space V^2 of ordered pairs of real numbers with the distance
    between points $P_1:(x_1, y_1)$ and $P_2:(x_2, y_2)$ defined by
    $$d(P_1, P_2) = [(x_1 - x_2)^2 + (y_1 - y_2)^2]^{1/2}.$$
    Let $m(P_1, P_2)$ be defined on V^2 by
    $$m(P_1, P_2) = |x_1 - x_2| + |y_1 - y_2|.$$
    Show that m is a metric for V^2. Show that if $x_1 \neq x_2$ and
    $y_1 \neq y_2$, then $S_m[P_1 P_2]$ is the set of all points in and
    on the euclidean rectangle with vertices (x_1, y_1), (x_2, y_1),

(x_2,y_2), (x_1,y_2). Show also that there are natural lines in the metric space (ν^2,m). If 0 is the origin, what does the circle $C_m(0,2)$ look like in terms of the euclidean plane?

6.   Suppose that it has been established that a possible metric m for a space s has the symmetry, positivity, and zero-identity properties.

   (i) Show that if A,B,C are elements of s which are not all distinct, then necessarily

   $$m(A,B) + m(B,C) \geqq m(A,C).$$

   (ii) Show that if $m(A,C) = \max \{m(A,B),m(B,C),m(C,A)\}$, then

   $$m(A,B) + m(B,C) \geqq m(A,C)$$

   implies both

   $$m(B,C) + m(C,A) \geqq m(B,A)$$

   and

   $$m(C,A) + m(A,B) \geqq m(C,B).$$

   (iii) Show that if $m(A,C) \geqq m(A,B)$ and $m(A,C) \geqq m(B,C)$, and if one of the equalities holds, then necessarily

   $$m(A,B) + m(B,C) \geqq m(A,C).$$

   (iv) Show, as a consequence of (i), (ii), and (iii), that to prove the triangle inequality with full generality, it suffices to prove that if A,B,C are three points of s such that $m(A,C) > m(A,B)$ and $m(A,C) > m(B,C)$, then

   $$m(A,B) + m(B,C) \geqq m(A,C).$$

7.   Show that in a metric space (s,m), $m(A,B) + m(B,C) > m(A,C)$ is implied by either m-<BCA>   or   m-<CAB> .

Topic II. A Spherical Metric

     In the euclidean space E^3, with metric d, let the sphere
with center O and radius a be denoted by S(O,a). If the points
of S(O,a) are taken to comprise a space, then (S(O,a),d) is a
metric space   (cf. T-I, Ex. 3). In E^3, the shortest path joining two points A,B, namely their geodesic connection, is the
segment S[AB]. However, if A,B are points of the space (S(O,a),d),
the path S[AB] is not contained in the space. Because the earth
is approximately a sphere, it is clearly a matter of practical
importance to know what is the geodesic connection of A and B in
(S(O,a),d), i.e. what is the shortest path from A to B on the
sphere. The answer, of course, is a matter of common knowledge,
though its proof does not belong to elementary mathematics. If
A and B are two points of the sphere which are not antipodal
(not collinear with O in E^3), then A,O,B determine a plane P(AOB).
The intersection of P(AOB) and S(O,a) is a circle C(O,a). All
circles on the sphere which have radius a are <u>great</u> <u>circles</u>.
The points A and B determine two arcs on C(O,a), and the minor
arc is the geodesic connection of A and B. If A and B are antipodal on the sphere then each plane, in the family of planes which
contain line L(AB), intersects S(O,a) in a great circle. Each
of the opposite semicircles determined by A and B, on any of
these circles, is a geodesic path from A to B.

     We will not attempt to prove that the great circle minor
arcs and semicircles are the geodesic paths in the metric space

(S(0,a),d). However, it is not difficult to show that the
euclidean lengths of such paths determine a metric for the
sphere, which we will denote by "s" rather than "m". Let the
radian measure of an angle $\measuredangle AOB$ be denoted by $(\measuredangle AOB)^{\#}$, so
$(\measuredangle AOB)^{\#} = \frac{\pi}{180} \measuredangle AOB^{\circ}$. It follows, by the definition of
radian measure, if two points A and B on the sphere are not
antipodal, then $a(\measuredangle AOB)^{\#}$ is the length of the great circle
minor arc joining them. The length of each great semicircle
is, of course, πa. These facts are the motivation for the
following definition of s(A,B), where A,B are points of the
sphere S(0,a):

(i) s(A,B) = s(B,A) = 0, if A = B;

(ii) s(A,B) = s(B,A) = $a(\measuredangle AOB)^{\#}$ if A,B are distinct and
not antipodal;

(iii) s(A,B) = s(B,A) = πa, if A and B are antipodal.

From the definition of s, it clearly has the symmetry,
positivity, and zero-identity properties of a metric. To show
that it also satisfies the triangle inequality, we need the
following theorem from euclidean solid geometry.

Theorem A.[*]

If R[OA), R[OB), and R[OC) are noncoplanar rays in E^3,
then each two angles in the set $\{\measuredangle AOB, \measuredangle BOC, \measuredangle COA\}$ have
measures whose sum is greater than that of the third.

Proof

It suffices to prove the inequality of the theorem for
the case in which the third angle has the maximum of the measures,
and we suppose that $\measuredangle AOC^{\circ} \geqq \measuredangle AOB^{\circ}$ and $\measuredangle AOC^{\circ} \geqq \measuredangle BOC^{\circ}$. If
the equality holds in either case, then clearly

$$\measuredangle AOB^{\circ} + \measuredangle BOC^{\circ} > \measuredangle AOC^{\circ},$$

[*] Book 11, Proposition 20, Euclid, T.L. Heath translation.

so we may suppose that $\angle AOC^0 > \angle AOB^0$ and $\angle AOC^0 > \angle BOC^0$.

In the plane P(AOC), there exists a ray R[OX) in the C-side of line L(OA) and such that $\angle AOX^0 = \angle AOB^0$. Because

$\angle AOC^0 > \angle AOX^0$, R(OX) $\subset$ In($\angle AOC$) and so R(OX) intersects S(AC) at a point D. Let E on R(OB) be the point such that $d(O,E) = d(O,D)$. Now $\triangle AOE \cong \triangle AOD$, by side-angle- side, so $d(A,E) = d(A,D)$. Therefore, from

$$d(A,E) + d(E,C) > d(A,C) = d(A,D) + d(D,C)$$

it follows that $d(E,C) > d(D,C)$. Since the triangles $\triangle EOC$ and $\triangle DOC$ have side S[OC] in common, and have congruent sides S[OE] and S[OD], $d(E,C) > d(D,C)$ implies that $\angle EOC^0 > \angle DOC^0$, (Th. 11, II-3). Because $\angle AOB^0 = \angle AOD^0$, and $\angle BOC^0 = \angle EOC^0 > \angle DOC^0$, it follows that

$$\angle AOB^0 + \angle BOC^0 > \angle AOD^0 + \angle DOC^0 = \angle AOC^0. \qquad \square$$

Now consider points A,B,C on the sphere S(O,a) and not coplanar with O. Then R[OA), R[OB), and R[OC) are not coplanar so Th. A implies that

$$\angle AOB^0 + \angle BOC^0 > \angle AOC^0.$$

Multiplying both sides of this inequality by $\frac{a\pi}{180}$ yields

$$a(\angle AOB)^{\#} + a(\angle BOC)^{\#} > a(\angle AOC)^{\#},$$

which is precisely the statement that

$$s(A,B) + s(B,C) > s(A,C).$$

Thus the strict triangle inequality holds for non-coplanar triples. The proof that $s(A,B) + s(B,C) \geq s(A,C)$ when A,B,C are coplanar with O is left as an exercise.

In the metric space $(S(O,a),s)$, as in the space $(V\ ,m)$ of T-I, Ex. 5, one can see that the lines and segments of a metric space may be such in name only. Their definitions express desirable properties, but, in themselves, these requirements are too weak to force lines and segments to have other reasonable properties. The space $(S(O,a),s)$ is an improvement over (V^2,m) in the sense that if A and B are two non-antipodal points then they are joined by a unique segment. However, if A and B are two antipodal points then every other point X satisfies s-<AXB>, thus the segment $S_s[AB]$ is the entire sphere. Even if A and B are not antipodal, it is not true that $X \in L_s(AB)$ implies that $L_s(AB)$ is either $L_s(AX)$ or $L_s(BX)$.

To explore spherical metrics a bit further we need some concepts from euclidean solid geometry which we review briefly. First, two intersecting planes P(ABC) and P(ABE) form four <u>dihedral angles</u> of intersection, each of which is a union of two closed half-planes. The notation ∢ D-AB-E indicates the dihedral angle which is the union of the two closed half-planes with L(AB) as a common edge and which pass through D and E respectively. The interior of ∢ D-AB-E is the intersection of two open half-spaces, namely the D-side of plane P(ABE) and the E-side of plane P(ABD). A plane which is perpendicular to L(AB) at a point P intersects ∢ D-AB-E in two closed rays, say R[PX) and

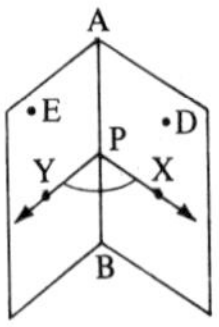

R[PY), and ∢ XPY is a <u>plane angle</u> of the dihedral angle. All the plane angles of a dihedral angle are congruent, and their common measure is defined to be the measure of the dihedral

angle. If L(AB) separates points D and D' in the plane P(ABD)
and separates E and E' in P(ABE), then ⊁D-AB-E and ⊁D'-AB-E'
are opposite dihedral angles and have equal measure.

If rays R[OA), R[OB), and R[OC) are non-coplanar in E^3,
the union of each of the angles, ⊁AOB, ⊁BOC, ⊁COA, with its
interior is a face of a trihedral angle which is defined as the
union of the three faces and denoted by ⊁ O-ABC. In addition
to its three face angles, a trihedral angle ⊁O-ABC has three
dihedral angles, namely ⊁B-OA-C, ⊁C-OB-A, ⊁A-OC-B, and the
intersection of their interiors is the interior of the trihedral angle.

It is characteristic of spherical metric geometries that
in such a space a triangle has an angle sum which exceeds 180°
by an amount called its excess. In model representations of
such geometries, an explanation for triangle excess can be
traced to the following theorem in euclidean solid geometry.

Theorem B.

The sum of the degree measures (the radian measures) of
the dihedral angles of a trihedral angle is greater than 180°
(greater than π).

Though we will not prove Th. B, a rather simple and
illuminating explanation of it can be obtained from the (non-
elementary) property that the surface area of a euclidean
sphere S(O,a) is $4\pi a^2$. First, if S[AB] is a diameter segment
of S(O,a) and points D,E on the sphere are such that A,B,D,E are
non-coplanar, then the dihedral angle ⊁ D-AB-E intersects S(O,a)
in two great semi-circles arc[ADB] and arc [AEB] . The points
of the sphere which are on or interior to ⊁D-AB-E form a

region called a "lune", which we denote by L[D-AB-E] corresponding to the dihedral angle of the lune. Let X and Y be
points on the sphere $S(0,1)$ such that the angle ∢XOY is a
plane angle of ∢D-AB-E, with X ∈ arc[ADB], with X' antipodal
to X, and with $\varphi = (∢XOY)^{\#}$. Then Y ∈ arc[AEB] and arc[XYX']
is a great semi-circle. If Y varies on arc(XYX'), it is clear
that the area of the lune L[X-AB-Y] = L[D-AB-E] varies in
direct proportion to φ. That is, doubling φ doubles the area,
tripling φ triples the area, and so on. As φ approaches π ,
the lune approaches a hemisphere with area $2\pi a^2$. Thus the

proportion
$$\frac{\text{ArL}[X-AB-Y]}{2\pi a^2} = \frac{\varphi}{\pi}$$

implies that the area of the lune is given by the formula
$$\text{ArL}[X-AB-Y] = 2a^2 \varphi. \tag{1}$$

Now, consider a trihedral angle ∢O-UVW in v^3. Let the
points at which the edge rays of the angle intersect a sphere
$S(0,a)$ be denoted by A,B,C respectively, so ∢O-UVW = ∢ O-ABC.
Designate by A',B',C' the points of the sphere antipodal to
A,B,C respectively and let the dihedral angles of ∢O-ABC
corresponding to the edge lines L(OA), L(OB), L(OC) have radian
measures α , β, γ respectively. Consider the hemisphere H in
the closed C-side of the plane through A,B,A',B'. If the nota-
tion T[ABC] denotes the region consisting of points on the
sphere which are on or interior to ∢O-ABC, then
$$H = T[ABC] \cup T[CBA'] \cup T[CAB'] \cup T[CA'B'] . \tag{2}$$

Since no two of the triangular regions intersect except along

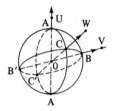

their boundaries, (2) implies that

$$\text{Ar } H = 2\pi a^2 = \text{ArT[ABC]} + \text{ArT[CBA']} + \text{ArT[CAB']} + \text{ArT[CA'B']}. \quad (3)$$

Because $T[ABC] \cup T[CBA'] = L[B\text{-}AA'\text{-}C]$, it follows from (1)
that

$$\text{ArT[ABC]} + \text{ArT[CBA']} = 2a^2 \alpha. \quad (4)$$

Similarly, $T[ABC] \cup T[CAB'] = L[C\text{-}BB'\text{-}A]$ implies that

$$\text{ArT[ABC]} + \text{ArT[CAB']} = 2a^2 \beta,$$

hence that

$$\text{ArT[CAB']} = 2a^2\beta - \text{ArT[ABC]} \quad (5)$$

The dihedral angles ⊰B-CC'-A and ⊰ B'-CC'-A' are opposite dihedral
angles, so both have measure γ. From $L[B'\text{-}CC'\text{-}A'] = T[CA'B'] \cup$
$T[A'B'C]$ it follows that

$$\text{ArT[CA'B']} = 2a^2\gamma - \text{ArT[A'B'C']}. \quad (6)$$

Because the reflection of E^3 in point O interchanges $T[ABC]$
and $T[A'B'C']$, these regions are congruent and have the same
area. Thus (6) can also be expressed in the form

$$\text{ArT[CA'B']} = 2a^2\gamma - \text{ArT[ABC]}. \quad (7)$$

Now, substitution of (4), (5), and (7) in (3) yields

$$2\pi a^2 = 2a^2\alpha + 2a^2\beta - \text{ArT[ABC]} + 2a^2\gamma - \text{ArT[ABC]},$$

which simplifies to the euclidean formula

$$\text{ArT[ABC]} = a^2(\alpha + \beta + \gamma - \pi). \quad (8)$$

Since the area of the region T[ABC] is a positive number,
(8) implies that $\alpha + \beta + \gamma > \pi$, which is the assertion of
Th. B.

To illustrate the usefulness of Th. B, we consider a metric
space which can be obtained from $(S(0,a),s)$ in such a way that
segments, rays, angles, and triangles can be defined in this
space, without ambiguity, following the pattern used in absolute
geometry. Let N be a point of $S(0,a)$ and let P_o denote the
euclidean plane which is perpendicular to L(ON) at 0 and which
intersects $S(0,a)$ in a great circle c_o . Let S^* be the intersection of $S(0,a)$ with the N-side of P_o , so $s*$ is an open
hemisphere with boundary c_o . Because $s^* \subset S(0,a)$, it follows
that (s^*,s) is a metric space (cf. T-I, Ex. 3). Moreover,
since c_o does not intersect s^* , no two points of s^* are antipodal on $S(0,a)$. Thus to each two points A,B in s^* there

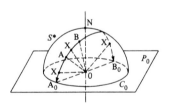

corresponds to a unique euclidean plane P(AOB) which intersects
c_o in a pair of points A_0, B_0 and intersects $s*$ in an open great
semicircle, arc(A_0B_0). If A_0, A, B, B_0, are successive in that order
on arc $[A_oB_o]$, then clearly $X \in$ arc (A_oA) implies s-<XAB>, and
$X \in$ arc(AB) implies s-<AXB>, and $X \in$ arc(BB_o) implies s-<ABX>.
Also, if $X \notin$ arc(A_oB_o) then no one of the points A,B,X is between

the other two. Thus arc(A_oB_o) is the line L_S(AB). In $s*$,
every open great semicircle of S(O,a) is an s-line; each two
points of $s*$ belong to exactly one s-line; every euclidean
plane through O, other than P_o, intersects $S*$ in a line of $(S*,s)$.

If two lines u,v of ($s*$,s) intersect at a point P, the
great circles of S(O,a) which carry u and v respectively intersect at P and at the point P' antipodal to P. Since P' $\notin$ $s*$,
it follows that if u,v intersect at P then they intersect only
at P. However, u and v need not intersect. Let u be the line
L_S(AB) in the diagram. Each plane, other than P_o, in the family
of planes which contain L(A_oB_o), intersects $s*$ in an s-line.
The carriers of these s-lines all intersect at A_o and B_o, which
are not points of $S*$, so no two of the s-lines intersect. If
we define two lines of $S*$ to be parallel if they do not intersect,
then there are infinitely many lines which are parallel to u.
Parallelism is weakly transitive. That is, if u,v,w are three
lines, then u $||$ v and v $||$ w imply that u $||$ w. It is also
interesting that the space $(s*,s)$ satisfies the Playfair
form of Euclid's parallel axiom (cf. Th. D, I-3).

In the space (S(O,a),s) there exist antipodal points whose
distance apart is πa, the maximum distance in the space. In
($s*$,s) the distance between two points is always less than πa,
but there exist pairs whose distance is less than πa by as little
as one pleases. The number πa is said to be the <u>least</u> <u>upper</u>
<u>bound</u> to the set of distances in ($S*$,s), so all lines of $S*$
may be regarded as having the same length πa.

If A and B are two points of $S*$, the line L_S(AB) is
carried by a great semicircle of S(O,a), say arc$[A_oB_o]$. If
B $\in$ arc(AB$_o$), then the points of arc$[AB_o)$ satisfy the definition
used in A^2 for the closed ray from A through B, so arc$[AB_o)$ **is**

$R_s[AB)$. If $C \notin L_s(AB)$, the union of $R_s[AB)$ and $R_s[AC)$ is the
angle $s-\sphericalangle BAC$, and the union of the closed segments $S_s[AB]$,
$S_s[BC]$, $S_s[CA]$ is the triangle $\triangle_s ABC$. A measure for $s-\sphericalangle BAC$
can be defined by the same pattern that was used in the Poincaré
model. On the sphere $S(0,a)$, the line $L_s(AB)$ is carried by a
great circle, and there is a euclidean ray $R[AX)$ in the plane
of this circle and tangent at A to the circular arc subtending
$\sphericalangle AOB$. Similarly, there is a ray $R[AY)$, in the plane of the
great circle carrying $L_s(AC)$, which is tangent at A to the
circular arc subtending $\sphericalangle AOC$. The degree measure of $s-\sphericalangle BAC$
is defined to be $\sphericalangle XAY^{\circ}$. Because $\sphericalangle XAY$ is a plane angle of
the dihedral angle $\sphericalangle B-OA-C$, it follows that the angle sum of
$\triangle_s ABC$ is the sum of the degree measures of the dihedral
angles of the trihedral angle $\sphericalangle O-ABC$. By Th. B, the angle
sum is greater than 180°.

Enough has been said to indicate that the geometry of
(S^*,s) presents some interesting analogies with ordinary
geometry. It also has some truly bizarre properties. To
list a few, if $P \neq N$, then the $s*$-circle $C_s(P, \frac{\pi a}{2})$ is a line.
If u is an S^*-line not through N, there exist points which have
no foot in u. Two s^*-circles may have a single intersection
point, even though each intersects the interior of the other.
A line through the center of a circle may intersect the circle
only once. Verification of these oddities is left to the
exercises.

Topic II Exercises

1. Complete the proof that $(S(0,a),s)$ is a metric space by
    showing that $s(A,B) + s(B,C) \geq s(A,C)$ when A,B,C are co-

planar with 0.

2. In the space $(S(0,a),s)$, show that if A and B are not anti-
    podal, and if ⧣AOB has radian measure α , then the length
    of the line $L_s(AB)$ is $a(2\pi - \alpha)$.

3. In the space ($s*$,s), show that for $0 < r < (\pi a)/2$ the
    circles $C_s(N,r)$ are also euclidean circles. If $P \neq N$,
    show that the circle $C_s(P,\pi a/2)$ is a line.

4. In $s*$, let u be a line which does not pass through N.
    The euclidean line perpendicular at 0 to the plane of u
    intersects $s*$ at a point Q. The plane which is perpen-
    dicular to P_o and which contains L(OQ) intersects C_o
    at a pair of points E_o, F_o and intersects $s*$ in the open
    semicircle arc(E_oF_o). One of the arcs, arc(QE_o) and
    arc(QF_o), is a minor arc.   Show that if P is a point of this
    minor arc then P has no foot in the line u.

5. If a euclidean angle ⧣AOB is revolved about the line L(OB),
    the ray R[OA) sweeps out a right circular cone. The line
    L(OB) is the axis of the cone and $\alpha = (⧣AOB)^{\#}$ is the
    angular measure of the cone. If P is a point of $s*$ and L(OP)
    is the axis of a right circular cone of angular measure α ,
    then all points of the cone which belong to $s*$ also belong
    to the S*-circle $C_s(P,a\alpha)$. Use this fact to explain how
    a line may pass through the center of a circle and intersect
    the circle only once. Explain also how two circles may
    intersect only once, even though each intersects the
    interior of the other.

6. Describe a location of points A and B in $s*$ for which
    the triangle Δ_s NAB obviously has an angle sum greater
    than 180°.

Topic III. Elliptic Geometry

The definition of a metric space places no restriction on
the nature of the space elements. That is, there is no restriction on what one chooses to call the "points" of the space.
This freedom can be used to obtain from a euclidean sphere a
model of elliptic plane geometry.

Consider a sphere in E^3 , for convenience a unit sphere
$S(0,1)$. For each point $A \in S(0,1)$, let A' denote the point
antipodal to A, so $(A')' = A$. Let the collection { X,X'}
of all antipodal pairs of $S(0,1)$ be taken as a space E^*,
with A* representing the point { A,A' } in E^*. Corresponding
to two points A* and B* in E^* , the euclidean lines $L(AA')$ and
$L(BB')$ intersect at O and form four angles of intersection.
The radian measure of the smallest of these angles is taken
as the elliptic distance between A* and B*, which we denote by
$e(A^*,B^*)$. The euclidean plane of the lines $L(AA')$ and $L(BB')$
intersects $S(0,1)$ in a great circle $C(0,1)$ and the pairs A,A'
and B,B' divide this circle into four non-overlapping arcs.

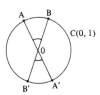

At least two of these are congruent, non-major arcs, say arc[AB]
and arc[A'B']. The quantity $e(A^*,B^*) = (\angle AOB)^\#$ is the common
length of these two shortest arcs.

More formally, corresponding to A* and B* in E^* , $e(A^*,B^*)$

is defined by:

(i) $e(A*,B*) = e(B*,A*) = 0$ if $A* = B*$;

(ii) $e(A*,B*) = \min \{ (\angle AOB)^{\#}, (\angle AOB')^{\#}\}$ if $A* \neq B*$.

Since $(\angle BOA)^{\#} = (\angle AOB)^{\#}$ and $(\angle BOA')^{\#} = (\angle AOB')^{\#}$, it follows

that $e(A*,B*) = e(B*,A*)$. Thus the measure e has the symmetry,

the positivity, and the zero-identity properties of a metric.

To establish the triangle inequality, we consider three

e-points (i.e. points of $E*$) $A*$, $B*$, $C*$ and suppose that

$$e(A*,C*) > e(A*,B*) \text{ and } e(A*,C*) > e(B*,C*). \qquad (1)$$

By. Ex. 6, T-I, the triangle inequality will be established

if we can show that these conditions imply that $e(A*,B*) +$

$e(B*,C*) \overset{\geq}{=} e(A*,C*)$.

Because $e(A*,C*)$ is one of the numbers $(\angle AOC)^{\#}$, $(\angle AOC')^{\#}$,

we may suppose the labels chosen so that

$$e(A*,C*) = (\angle AOC)^{\#}, \qquad (2)$$

and hence that

$$(\angle AOC')^{\#} = (\angle A'OC)^{\#} \overset{\geq}{=} (\angle AOC)^{\#} = (\angle A'OC')^{\#}. \qquad (3)$$

Suppose, first, that $A*,B*,C*$ are coplanar and hence belong

to a great circle $C(0,1)$. If $e(A*,C*) = (\angle AOC)^{\#} = \frac{\pi}{2}$, then

$B*$ must have one of the points B,B', say B, on the $90°$ arc(AC)

or else on the $90°$ arc(AC'). If $B \in$ arc(AC), then $(\angle AOB)^{\#} +$

$(\angle BOC)^{\#} = \frac{\pi}{2}$.

If B ∈ arc(AC'), then $(\sphericalangle AOB)^{\#} + (\sphericalangle BOC')^{\#} = \frac{\pi}{2}$. Thus, in
either case,

$$e(A^*,B^*) + e(B^*,C^*) = e(A^*,C^*). \qquad (4)$$

If $e(A^*,C^*) = (\sphericalangle AOC)^{\#} < \frac{\pi}{2}$, let D on C(0,1) be such that
C is interior to the right angle $\sphericalangle AOD$ and let E on C(0,1) be
such that A is interior to the right angle $\sphericalangle COE$. If point X
belongs to the minor arc(CD), then C is between A and X on

arc(ACX), which is a non-major arc. Therefore e-< A*C*X* >,
which implies that $e(A^*,X^*) > e(A^*,C^*)$, and hence, from (1),
that X* ≠ B*. Similarly, if X belongs to minor arc (AE], then
A is between C and X on the non-major arc(CAX), hence e-<X*A*C* >,
which implies that $e(X^*,C^*) > e(C^*,A^*)$, and so, by (1), X* ≠
B*. Thus B* must have one member of its antipodal pair B,B',
say B, interior to $\sphericalangle AOC$ or interior to $\sphericalangle DOE'$. If B is interior to $\sphericalangle AOC$, which is acute, then clearly e-<A*B*C* >. If B ∈
In($\sphericalangle DOE'$), then B is also interior to the right angle $\sphericalangle A'OD$,
and so $e(A^*,B^*) = (\sphericalangle A'OB)^{\#}$. But B ∈ In($\sphericalangle DOE'$) also implies
that B is interior to the right angle $\sphericalangle E'OC$, and hence that
$e(B^*,C^*) = (\sphericalangle BOC)^{\#}$. Therefore,

$$e(A^*,B^*)+e(B^*,C^*)=(\sphericalangle A'OB)^{\#}+(\sphericalangle BOC)^{\#}=(\sphericalangle A'OC)^{\#} > \frac{\pi}{2} > e(A^*,C^*).$$

Thus, in both cases,

$$e(A^*,B^*) + e(B^*,C^*) \overset{\geq}{=} e(A^*,C^*). \qquad (5)$$

From (4) and (5), the triangle inequality holds for all coplanar triples.

Next, suppose that A*,B*,C* are not coplanar. The two angles, ∢AOB and ∢BOC must satisfy exactly one of the following three size relations: both of them are non-obtuse; only one of them is non-obtuse; both are obtuse. We consider these possibilities separately.

Case 1

Both ∢AOB and ∢BOC are non-obtuse. Therefore, by definition, $(∢AOB)^{\#} = e(A*,B*)$ and $(∢BOC)^{\#} = e(B*,C*)$. By Th. A, T-II,

$$(∢AOB)^{\#} + (∢BOC)^{\#} > (∢AOC)^{\#}$$

hence

$$e(A*,B*) + e(B*,C*) > e(A*,C*). \qquad (6)$$

Case 2

Exactly one of the angles, ∢AOB and ∢BOC, is non-obtuse. Suppose, first, that $(∢AOB)^{\#} \leq \frac{\pi}{2}$ and that $(∢BOC)^{\#} > \frac{\pi}{2}$. Now, $(∢AOB)^{\#} = e(A*,B*)$ and $(∢BOC')^{\#} = e(B*,C*)$. By Th. A, T-II,

$$(∢AOB)^{\#} + (∢BOC')^{\#} > (∢AOC')^{\#}. \qquad (7)$$

From (2) and (3), $(∢AOC')^{\#} \geq e(A*,C*)$, so (7) implies that

$$e(A*,B*) + e(B*,C*) > e(A*,C*). \qquad (8)$$

By an entirely similar argument, (8) holds when it is ∢AOB which is obtuse and ∢BOC which is non-obtuse.

Case 3

Angles ∢AOB and ∢BOC are both obtuse. Now, $(∢AOB')^{\#} = e(A*,B*)$ and $(∢B'OC)^{\#} = e(B*,C*)$. By Th. A, T-II,

$$(\sphericalangle AOB')^{\#} + (\sphericalangle B'OC)^{\#} > (\sphericalangle AOC)^{\#},$$

hence

$$e(A^*, B^*) + e(B^*, C^*) > e(A^*, C^*). \qquad (9)$$

Equations (6), (8) and (9) complete the proof that e satisfies the triangle inequality and hence that (E^*, e) is a metric space. These equations also show that the strict triangle inequality holds for any three e-points which are not coplanar in E^3.

In proving the triangle inequality, we made use of the betweeness of points in E^*, as betweeness was defined in Topic I. However, we now introduce a definition for a line in E^* which differs from the definition in T-I.

Definition. (Line in E^*) If A* and B* are two points of E^*, the euclidean plane of L(AA') and L(BB') intersects the sphere S(O,1) in a great circle C(O,1). The set of all antipodal pairs {X,X'} contained in C(O,1) is the elliptic line, or e-line, of A* and B*, and we denote it by $L_e(A^*, B^*)$.

To see that the lines of E^*, as just defined, differ from those defined in T-I, consider a great circle C(O,1) on the sphere S(O,1), and let A,B,C,A',B',C' be the consecutive vertices of a regular hexagon inscribed in C(O,1). The pairs A,A' and B,B', and C,C' are antipodal, so, by definition, C(O,1) = $L_e(A^*B^*) = L_e(B^*C^*) = L_e(C^*A^*)$. However, because all the consecutive minor arcs, subtended by the sides of the hexagon, are $60°$ arcs, $e(A^*, B^*) = e(B^*, C^*) = e(C^*, A^*)$. Thus no one of the points A*,B*, C* is between the other two.

An e-line, corresponding to a great circle on S(O,1), has some properties which are normally associated with lines and others which are normally associated with circles. Among the ordinary properties, for example, two e-points on C(O,1) belong

to this great circle and no other, so an e-line is uniquely
determined by any two of its points. Also, each two points of
E* belong to exactly one line. Because each two great circles
on S(0,1) intersect at a pair of antipodal points, each two e-
lines intersect at a single e-point, and this is a fundamental
property of the elliptic plane.

The sense in which an e-line is like a euclidean circle
is apparent in the example of a regular euclidean hexagon A,B,
C,A',B',C' inscribed in a great circle C(0,1). On the line

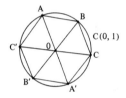

$L_e(A*B*)$, there are clearly two paths from A* to B*. One is
the set of points X* such that X* = A*, X* = B*, or e-<A*X*B*>,
namely the segment $S_e[A*B*]$, and this is the set of antipodal
pairs of S(0,1) in the union of the minor arcs, arc[AB] ∪
arc[A'B']. The second path from A* to B* in the line is that
via C*, and this is the set of antipodal pairs of C(0,1) in the
union arc[AC'B'] ∪ arc[BCA']. Since $e(A*,B*) = \frac{\pi}{3}$, it is natural
to define this as the length of $S_e[A*B*]$. The second path is
also represented by $S_e[A*C*] ∪ S_e[C*B*]$, and since these segments intersect only at C*, it is natural to regard the length
of this path as $\frac{2\pi}{3}$, and to regard the line $L_e(A*B*)$ as having
length π.

When $0 < e(A*,B*) < \frac{\pi}{2}$, the e-points between A* and B*
uniquely determine the open segment $S_e(A*B*)$. In the same way,
if two points A,B on a unit euclidean circle have arc distance

less than π, then the minor arc(AB) is uniquely determined.
But just as A,B on the circle may be antipodal, and determine
two semicircles, so A* and B* may be <u>opposite</u> <u>points</u> on
$L_e(A^*B^*)$ in the sense that $e(A^*,B^*) = \frac{\pi}{2}$. In this case, the
locus of points X* which satisfy the relation

$$e(A^*,X^*) + e(X^*,B^*) = e(A^*,B^*)$$

is precisely the line $L_e(A^*B^*)$. If C* on the line is distinct
from A* and B*, then the natural path from A* to B* via C* is
$S_e[A^*C^*] \cup S_e[C^*B^*]$. If D* on the line is not in this path,
the the path from A* to B* via D*, represented by $S_e[A^*D^*] \cup$
$S_e[D^*B^*]$, is a second path, and both paths have length $\pi/2$.
To distinguish the sets, one can define $S_e[A^*C^*B^*]$ to mean
that A*,B*,C* are three collinear points, that $e(A^*,B^*) = \pi/2$,
and that $S_e[A^*C^*B^*] = S_e[A^*C^*] \cup S_e[C^*B^*]$. Then if $D^* \in L_e(A^*B^*)$,
and $D^* \notin S_e[A^*C^*B^*]$, it follows that $S_e[A^*D^*B^*]$ is the second
closed segment joining A* and B*. Naturally enough, the open
and closed segments $S_e(A^*C^*B^*)$ and $S_e[A^*C^*B^*]$ are also called open
and closed half-lines.

Not only does an e-line have some circle properties, there
is a sense in which it is a circle. Corresponding to P* and
$0 < r < \pi/2$, the e-circle $C_e(P^*,r) = \{X^*: e(P^*,X^*) = r\}$ is
represented on $S(0,1)$ by a pair of congruent euclidean small
circles. These lie in euclidean planes which are perpendicular
to $L(PP')$ and symmetric with respect to 0. If P and P' are
thought of as north and south poles on $S(0,1)$, then $C_e(P^*,r)$
is the union of two circles of latitude which have the same
north and south latitude. As r increases, both euclidean
circles move toward the equator and when $r = \pi/2$ the circle
$C_e(P^*, \frac{\pi}{2})$ is an e-line, namely the line represented by the
equatorial great circle. On each e-line u in the pencil $P_e(P^*)$

there is a point X^* opposite to P^* and the e-line $t = C_e(P^*,X^*)$
is the locus of all such points. Line t is called the <u>polar</u>

<u>line</u> to P^*, and P^* is the <u>pole</u> of line t. The pole-polar re-
lations of points and lines play an important role in elliptic
geometry. It is easy to see, for example, that a point Q^* is
on the line t which is polar to P^* if and only if P^* is on the
line s which is polar to Q^*. The line $L_e(P^*Q^*)$ is then the
polar line to $\{R^*\} = s \cap t$.

An angle in (E^*,e) may be defined as the union of two
non-collinear closed half-lines which have a common endpoint.
However, care must be exercised in the representation of an
angle if ambiguity is to be avoided. Consider three non-
collinear e-points A^*,B^*,C^*, such that neither B^* or C^* is on
the polar line to A^*. On $u = L_e(A^*,B^*)$ there is a point B_1^*
opposite to A^*, and $S_e[A^*B^*B_1^*]$ is the closed half-line of u,
with endpoint A^*, which contains B^*. Similarly, on $v=L_e(A^*C^*)$
there is a unique closed half-line $S_e[A^*C^*C_1^*]$, with endpoint
A^* and containing C^*. These closed half-lines on u and v
respectively are determined by A^* and B^* and by A_1^* and C_1^*, and
the union of the two half-lines is the e-angle represented by

e- ⧊B*A*C*.

If either B* or C* is on the polar line to A*, then the
notation e- ⧊B*A*C* is ambiguous. For example, if e(A*,B*) =
π/2, then on u = L$_e$(A*B*) there exist two closed half-lines,
S$_e$[A*P*B*] and S$_e$[A*Q*B*], with only the points A*,B* in
common, and one must specify which of the two half-lines is
intended in the union which represents the angle.

When the half-lines, or arms, of an e-angle have been
clearly specified, a measure for the angle can be defined by the
same sort of device used in the Poincaré model. Consider the
angle e-⧊B*A*C*, where neither B* or C* is on the polar line
to A*. On the sphere S(0,1), the antipodal pairs representing
A*,B*, and C* can always be labeled so that ⧊AOB and ⧊AOC are
acute. The line u = L$_e$(A*B*) is a great circle C$_u$(0,1) on the
sphere, and in the plane of this circle there is a euclidean
ray R[AX) which is tangent at A to the arc subtending ⧊AOB.
Similarly, the line v = L$_e$(A*C*) is a great circle C$_v$(0,1) on
the sphere, and in the plane of this circle there is a euclidean
ray R[AY) which is tangent at A to the arc subtending ⧊AOC.
The euclidean measure of ⧊XAY is defined to be the measure of
e- ⧊B*A*C*. This is also, of course, the measure of one of
the dihedral angles formed by the euclidean planes which carry
u and v respectively. The measure of the angles at which the
euclidean lines L(AX) and L(AY) intersect provide measure for
the angles at which the e-lines u and v intersect. In particular, u and v are perpendicular when the u and v planes are
perpendicular.

When an angle is defined directly as a union of half-lines,
say e-⟶B*A*C* = S$_e$[A*D*B*] ∪ S$_e$[A*E*C*], the antipodal pairs
representing A*, D*, and E* can be labeled so that ⧊AOD and

∡ AOE are acute. As before, the measure of e-∡ B*A*C* is
defined to be the measure of a euclidean angle ∢XAY, where
R[AX) is tangent at A to a circular arc subtending ∡AOD and
R[AY) is tangent at A to a circular arc subtending ∡AOE.

The same carefulness necessary in the representation of
elliptic angles must also be used in the representation of
elliptic triangles. If points A*, B*, C* are non-collinear
points such that no one of them is on the polar line to another,
then there is no difficulty about the meaning of the e-triangle
$\Delta_e A*B*C*$.   Each two of the three points A*,B*,C* determines
a unique closed segment, and the triangle is the union of these
three sides. The angles e-∡ A*B*C*, e- ∡ B*C*A*, and e- ∡ C*A*B*
are also well defined and are the angles of the triangle.
However, if two of the vertices are opposite points on the e-
line they determine, for example if e(A*,B*) = $\pi/2$, then one
must specify which of the two closed half-lines with endpoints
A*,B* is intended to be the "side opposite to C*". Once the
side opposite to each vertex is well defined, then the angle
of the triangle at A*, for example, is defined by the closed
half-lines with endpoint A* and which contain the sides opposite
to B* and C* respectively.

Though we will not give the proof here, it can be shown
that in every e-triangle the (degree) angle sum is greater than
180°. One instance of this is particularly obvious. Consider
three points A*, B*, C* such that the euclidean lines L(AA'),
L(BB'), and L(CC') are pairwise perpendicular. Then, clearly,
each two of the lines L_e(A*B*), L_e(B*C*), L_e(C*A*) are perpendicular. And however X*, Y*, Z* are selected, the e-triangle
which is the union of the three closed half-lines, S_e[A*X*B*] ,
S_e[B*Y*C*], and S_e[C*Z*A*] , has three right angles and an

angle sum of 270°.

As mentioned earlier, there are axiom systems for elliptic
plane geometry. It can be proved that, with proper interpretation, the points, segments, lines, and angles in (E^*,e)
satisfy those axioms and hence that (E^*,e) is a model of the
elliptic plane. The foregoing discussion simply gives some
indication of how such a proof could be obtained.

Topic III Exercises

1. Let P, Q, R, denote three points of $S(0,1)$ such that P,
    Q,R,0 are non-coplanar and $(\maltese POQ)\# = (\maltese QOR)\# = (\maltese ROP)\# = \pi/2$. Let A, B, C be the midpoints of the 90° arcs sub-
    tending these right angles. Show that the length of the
    minor arc[AC] is greater than $\pi/4$. Thus show that the
    segment $S_e[A*C*]$ joining the midpoints of sides $S_e[P*A*Q*]$
    and $S_e[P*C*R*]$ has length greater than half of the third
    side $S_e[P*B*Q*]$ in the triangle $\Delta_e[P*Q*R*]$.

2. Describe and picture a Saccheri quadrilateral in (E^*,e)
    in which the summit is shorter than the base.

3. Describe the points of $S(0,1)$ which form the region interior
    to the circle $C_e(P*,\pi/3)$ and also those which form the
    exterior region.

4. Describe the family of e-circles which are tangent to
    $C_e(P*,\pi/3)$ and also tangent to the line polar to P*. What
    is the radius of such a circle? If c_1 is a circle in
    this family, how many other e-circles in the family are
    also tangent to c_1?

Topic IV. Barbilian Geometries, the Cross Ratio Metric

As a final topic, we consider a class of metric spaces imbedded in the euclidean plane E^2, which were defined by D. Barbilian in 1934. As we shall see, the logarithm cross ratio metric of the Poincaré model arises here in a natural way, and the model for H^2 appears as a special Barbilian geometry.

In the euclidean plane, let c denote a simple closed curve which divides the plane into a region interior to c and a region exterior to c. The points of the region interior to c are the points of a Barbilian space B^2. Corresponding to two points A,B in B^2, consider the ratio $d(X,A)/d(X,B)$, where X is a point of c.

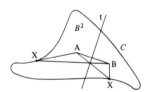

As X varies on c, the number $d(X,A)/d(X,B)$ varies and takes on a maximum value for at least one point P* on c (a geometric argument for this will be given presently). Let $r(A,B)$ be defined by

$$r(A,B) = \max_{X \in c} \{\frac{d(X,A)}{d(X,B)}\} = \frac{d(P^*,A)}{d(P^*,B)}.$$

Consider the line t which is the perpendicular bisector of the segment S[AB]. Since $d(X,A)/d(X,B) = 1$ for all $X \in t$ and $d(X,A)/d(X,B) < 1$ for all $X \in$ A-side of t, then P* must be in the B-side of t, hence $r(A,B) > 1$. In the case for which A = B, $d(A,X)/d(X,B) = 1$ for all $X \in c$, so in this case $r(A,B) = 1$.

From the foregoing properties, $\log r(A,B) > 0$ if A ≠ B and $\log r(A,B) = 0$ if and only if A = B, which are relations which suggest that $\log r(A,B)$ might be a metric. However, since

$$\max_{X \in c} \{\frac{d(X,A)}{d(X,B)}\} \qquad \text{and} \qquad \max_{X \in c} \{\frac{d(X,B)}{d(X,A)}\}$$

need not be equal, log $r(A,B)$ and log $r(B,A)$ need not be equal, so the symmetry property is not satisfied. But consider the product $r(A,B)r(B,A)$, which is symmetric in A and B. If $A \neq B$, then $r(A,B) > 1$ and $r(B,A) > 1$ imply that $\log[r(A,B)r(B,A)] > 0$. Also, since $r(A,B) = 1$ if and only if $A = B$, and $r(B,A) = 1$ if and only if $A = B$, it follows that $\log[r(A,B)r(B,A)] = 0$ if and only if $A = B$. Thus a Barbilian metric b defined by

$$b(A,B) = \log[r(A,B)r(B,A)], \quad A,B \in B^2 \tag{1}$$

has the positivity, zero-identity and symmetry properties. If P^* and Q^* are maximizing points for $r(A,B)$ and $r(B,A)$ respectively, then

$$b(A,B) = \log \left[\frac{d(P^*,A)}{d(P^*,B)} \frac{d(O^*,B)}{d(Q^*,A)}\right] \tag{2}$$

which is the logarithm of a cross ratio.

It is still to be shown that b satisfies the triangle inequality. To this end, let A,B,C denote points of B^2 and let X_O be a point of c. From the definitions of $r(A,B)$ and $r(B,C)$ it follows that

$$r(A,B) \geq d(X_O,A)/d(X_O,B) \text{ and } r(B,C) \geq d(X_O,B)/d(X_O,C),$$

hence

$$r(A,B)r(B,C) \geq d(X_O,A)/d(X_O,C). \tag{3}$$

Because (3) holds for every point X_O in c, it follows that

$$r(A,B)r(B,C) \geq r(A,C). \tag{4}$$

By the same reasoning,

$$r(B,A)r(C,B) \geq r(C,A), \tag{5}$$

and from (4) and (5),

$$\log[r(A,B)r(B,C)r(B,A)r(C,B)] \geq \log[r(A,C)r(C,A)]. \tag{6}$$

By the properties of logarithms, (6) implies that

$$\log[r(A,B)r(B,A)] + \log[r(B,C)r(C,B)] \geq \log[r(A,C)r(C,A)] \tag{7}$$

which is the statement that

$$b(A,B) + b(B,C) \geq b(A,C). \tag{8}$$

Thus (B^2,b) is a metric space.

The determination of the maximizing points P*, Q* in (2) involves some classical elementary geometry. As background, consider the following problem first solved by Apollonius (260?-200?, B.C.). Given two points A, B in E^2, and a number $k > 1$, find the locus L of all points X in E^2 such that $d(X,A)/d(X,B) = k$. To locate the points of L on the line $L(AB)$, let $R(AC)$ and $R(AD)$ denote rays not contained in $L(AB)$. On $R(AC)$ there exist points C_1 , C_2 such that $d(A,C_1) = k$, $d(C_1,C_2) = 1$, and $<AC_1C_2>$. On $R(AD)$ there exist points D_1, D_2 such that $d(A,D_1) = k$, $d(D_1,D_2) = 1$, and $<AD_2D_1>$. The line through C_1 and parallel to $L(C_2B)$ intersects $L(AB)$ at a point Y, and $d(A,Y)/d(Y,B) = d(A,C_1)/d(C_1,C_2) = k/1$ implies that $Y \in L$, and $< AC_1C_2 >$ implies $<AYB>$. The line through D_1 and parallel to

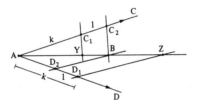

$L(D_2B)$ intersects $L(AB)$ at a point Z, and $d(Z,A)/d(Z,B) = d(D_1,A)/d(D_1,D_2) = k/1$ implies that $Z \in L$, and $< AD_2D_1 >$ implies $< ABZ >$.

The circle c_k which has $S[YZ]$ as a diameter segment is known as the "circle of Apollonius", and is the desired locus L.   To see this, consider any point P which is on the circle c_k and is distinct from Y and Z. The lines through B and parallel to $L(PY)$ and $L(PZ)$ respectively intersect $L(AP)$ at points E and F respectively. The angle $\not{}YPZ$, inscribed in a semicircle of c_k, is a right angle.

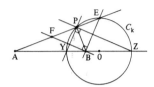

Thus the parallelogram with diagonal S[BP] and opposite sides on
L(BE) and L(PY) respectively is a rectangle, and ∡EBF is a
right angle. Because L(YP) || L(BE),

$$\frac{d(P,A)}{d(P,E)} = \frac{d(Y,A)}{d(Y,B)} = k, \tag{9}$$

and because L(ZP) || L(BF),

$$\frac{d(P,A)}{d(P,F)} = \frac{d(Z,A)}{d(Z,B)} = k. \tag{10}$$

From (9) and (10), it follows that d(P,E) = d(P,F). Therefore
P is the midpoint of hypotenuse S[EF] in the right triangle
△ EBF.   Thus P is the center of the circumcircle of △ EBF,
hence

$$d(P,E) = d(P,B) = d(P,F). \tag{11}$$

The combination of (9) and (11), or (10) and (11), shows that
d(P,A)/d(P,B) = k. Therefore, P ∈ L, which implies that
$C_k \subset L$.   The converse, namely that if P ∈ L then P ∈ c_k
is left as an exercise.

Let O denote the center of c_k and let a = d(O,Y) = d(O,Z).
From (9) and (10), <BOZ> (c f. Ex. 2), so d(Y,A) = d(O,A) -
d(O,Y) = d(O,A) - a, and d(Y,B) = d(O,Y) - d(O,B) = a - d(O,B).
Thus

$$k = \frac{d(Y,A)}{d(Y,B)} = \frac{d(O,A) - a}{a - d(O,B)} . \tag{12}$$

Similarly, d(Z,A) = d(Z,0) + d(O,A) = a + d(O,A), and d(Z,B) =
d(Z,0) + d(O,B) = a + d(O,B), and so

$$k = \frac{d(Z,A)}{d(Z,B)} = \frac{a + d(O,A)}{a + d(O,B)} . \tag{13}$$

Equations (12) and (13) imply that

$$\frac{d(O,A) - a}{a - d(O,B)} = \frac{a + d(O,A)}{a + d(O,B)},$$

which simplifies to

$$d(O,A)d(O,B) = a^2. \tag{14}$$

Thus A and B are inverses with respect to the circle c_k. From Cor. 11, IV-2, it follows that if c_M is the circle with diameter S[AB] and center M, then Y and Z are inverses with respect to c_M.

It is now easy to see how the Apollonian circles c_k vary as k varies, with k > 1. Under inversion in c_M, the open segment S(BM) inverts to the open ray opposite to R(BM). Each point Y of S(BM) determine a number k = d(Y,A)/d(Y,B) > 1, and conversely, and c_k is the circle with diameter S[YZ] , where Z is the inverse of Y. If <MY$_1$Y$_2$> then <BZ$_2$Z$_1$>. The number k_1 = d(Y$_1$,A)/d(Y$_1$,B) is less than k_2 = d(Y$_2$,A)/d(Y$_2$,B), and c_{k_1} contains c_{k_2} in its interior. As Y approaches B, or equivalently as k increases, Z also approaches B, and the decreasingly nested

circles approach B as a limit. As Y approaches M, or equivalently as k decreases toward 1, the distance d(B,Z) increases without bound, and the increasingly nested circles approach, as a limit, the line t which is the perpendicular bisector of S[AB] . The union of all the circles c_k, for k > 1, is except for B the B-side of line t.

Returning to the subject of Barbilian geometry, let A,B denote two points of a Barbilian space (B^2,b), and let t be the line perpendicular to S[AB] at the midpoint M. If Y$_1$ on S(MB) is sufficiently near to B, the Apollonian circle c_{k_1}, corresponding to k_1 = d(Y$_1$,A)/d(Y$_1$,B), is contained in B^2 and does not intersect the boundary c. As Y moves toward M, then $d(Y,A)/d(Y,B)$ decreases but the circles c_k increase in size until one of them, say c_{k*}, makes first contact with the boundary c,

possibly at more than one point. Each contact point P* of c_{k*} with c is a maximizing point for $r(A,B)$. For if X on c is in the B-side of t but is not on c_{k*}, then X is exterior to c_{k*} and belongs to some Apollonian circle c_k larger than c_{k*}. Therefore $d(X,A)/d(X,B) = k < k* = d(P*,A)/d(P*,B)$. The maximizing point Q* for $r(B,A)$ is determined in a similar way, using Apollonian circles c_k in the A-side of t, where $k = d(Y,B)/d(Y,A)$ and $Y \in S(MA)$.

Consider now the case in which a Barbilian space is the interior of a euclidean circle c with center I. If two points A,B of B^2 are collinear with I, let Z be the intersection of c with the ray R(AB) and let c_M denote the circle with diameter S[AB] and center M. If Y is the inverse of Z with respect to c_M, then the circle with diameter S[YZ] is the Apollonian circle c_k for $k = d(Z,A)/d(Z,B) = d(Y,A)/d(Y,B)$. Since c_k is tangent to c, it follows that Z = P* is the maximizing point for $r(A,B)$. Similarly, the other intersection of c and L(AB) is the point Q* which is the maximizing point of

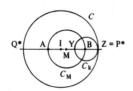

r(B,A). Thus, in this case, b(A,B) is precisely the hyperbolic distance h(A,B) in the Poincaré model with boundary c .

If A and B in B^2 are not collinear with I, let M again be the midpoint of S[AB]. For Y variable on S(MB) and k = d(Y,A)/d(Y,B) the Apollonian circle c_k of first contact with c is tangent to c at a point P* not on L(AB). Let $c*$ denote the circle determined by A, B, and P*. As was proved earlier, A

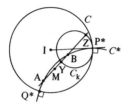

and B are inverse points with respect to c_k. Therefore $c*$, which passes through A and B, is orthogonal to c_k at P*. Since c_k is tangent to c at P*, it follows that $c*$ is also orthogonal to c at P*. Thus the maximizing point for r(A,B) is an intersection point of c with a circle through A,B and orthogonal to c. By symmetry, this must also be true for the maximizing point of r(B,A). But C* is the only circle through A and B which is orthogonal to c . Thus the second intersection of $c*$ and c is the maximizing point Q* for r(B,A). Thus, in all cases, b(A,B) is the hyperbolic distance h(A.B) of the Poincaré model, which establishes that in fact (B^2,b) is the Poincaré model of hyperbolic geometry.

Topic IV Exercises

1. Given two points A, B in the euclidean plane, and a number
    k > 1, it was shown that there exist points Y and Z on
    L(AB) such that
    $$d(Y,A)/d(Y,B) = d(Z,A)/d(Z,B) = k,$$
    with < AYB > and <ABZ> . Prove that if point P, not
    on L(AB), is such that $d(P,A)/d(P,B) = k$, then P belongs
    to the circle with S[YZ] as a diameter segment. Show also
    that L(PY) and L(PZ) are the internal and external bisec-
    tors of ≮APB in ΔAPB.

2. Use equations (9) and (10), and the fact that d(Z,A) >
    d(Y,A), to show that d(Z,B) > d(Y,B). How does this imply
    that the midpoint O of S[YZ] is between B and Z?

3. In the argument preceding equation (8), which established
    the triangle inequality for the metric b, show that one
    obtains the equality in (8) if and only if one has equality
    at all the preceding stages. Thus show that a Barbilian
    point B is between two others, A and C, if and only if there
    exist points P* and Q* on c such that P* is a maximizing
    point for r(A,B), r(B,C) and r(A,C) while Q* is a maximizing
    point for r(B,A), r(C,B) and r(C,A).

4. Let $C(O_1,a)$ and $C(O_2,a)$ be two euclidean circles which are
    orthogonal at points A and B. Let the union of the inter-
    iors of the circles be taken as a space (B^2,b). Show that
    the euclidean segment S(AB) is a natural line in the space
    (B^2,b).

Index

(Numbers refer to pages)

-V-